Military collectables

An international directory of twentieth-century militaria

Military

collectables

An international directory of twentieth-century militaria

Chief consultant: Joe Lyndhurst

MAGNA BOOKS

A Salamander Book

Published by
Magna Books
Magna Road
Wigston
Leicester
LE8 2ZH

ISBN 0 948509 21 X

Editor:
Richard O'Neill

Designer:
Barry Savage

Photography:
Terry Dilliway
© Salamander Books Ltd.

Filmset:
Modern Text Typesetting Ltd.

Colour reproduction:
Bantam Litho Ltd.

Printed by:
Henri Proost et Cie, Turnhout, Belgium

Acknowledgements

In the preparation of this book we received the most generous assistance and advice from militaria collectors and museum directors and staff in the United Kingdom and the Federal Republic of Germany. Private collections were opened to us for research and photography; museums allowed us to pillage their displays and ransack their store-rooms in search of material. Without the unstinted help of the individuals and organisations listed below, this record of the full range of 20th-century militaria could not have been assembled.

Primarily, our thanks are due to **Joe Lyndhurst** of the Warnham War Museum, West Sussex, the chief consultant on this project and the provider, from his truly remarkable collection, of much of the material photographed. We also thank **Mrs Yvonne Lyndhurst,** both for her great logistical achievements on behalf of our photographic team and for allowing photography of items from her own collection, and **Michael Holmes** of the Warnham War Museum.

Detailed acknowledgement of the collections from which material was drawn for photography is made in the Picture Credits. Here, we acknowledge the help given by:

Mr Tony Bradley,
Chingford, London
Mr F. Caldwell/Mr H.J. Woodend,
Pattern Room Curator, Royal Small Arms Factory, Enfield Lock, Middlesex
Mr Bob Cayley,
Brighton
Commanding Officer and Staff of the Wehrgeschichtliches Museum, Rastatt-Schloss, Baden, Federal Republic of Germany: in particular,
Major d.R. Henning Volle
(Medals and Metal Insignia);
Fräulein Sabina Hermes
(Printed Material);
Oberleutnant Udo U. Lander
(Weapons);
Major Christian-Wilhelm von Prittwitz und Gaffron
(Uniforms);
Herr Harald Rosmanitz
(Cloth Insignia);
Herr Ulrich Schiers
(Headgear)
Mr Malcolm Fisher,
"Regimentals", Islington, London
Major (Retd) M.J. Fogwell/Mr F. Davie,
Weapons Museum, School of Infantry, Warminster, Wiltshire
Mr Terry Gander,
Billingshurst, Sussex
Mr Gordon Gardiner,
Sotheby's, Pulborough, Sussex, and, for their assistance with the loan of edged weapons for photography,
Messrs Eric Campion, David Chester, Trevor Edmonds, Peter Gasnier, Kevin Holmes, Graham Lay, Alistair Morris and **Andrew Todman**
Mr A. Litherland,
Spink & Son, London
Miss Sylvia March,
Brighton
Mr Richard Mayne,
Museum of Nazi German Equipment & Occupation Relics, St Peter's Bunker, Jersey, Channel Islands
Lieutenant-Colonel D.M.O. Miller,
Twickenham, London
Lieutenant-Colonel (Retd) J.B.R. Nicholson,
Ivychurch, Kent
Miss Margaret E. Nobbs,
Leicester
Mr Mike Priest,
Priest Firearms & Militaria, Bromley, Kent
Dr A. Sasse,
Bad Oeynhausen, Federal Republic of Germany
Mr J.D. Sheen,
Crawley Coins & Medals, Crawley, Sussex
Mr Tony Walker/Mrs Vannessa Walker,
Lewisham, London
Mr Doug Warneck,
Gillingham, Kent
Mr Robert Wilkinson-Latham,
Redenham, Hampshire

Contents

Foreword

My lifelong interest in militaria began when I was a child and, as far as I can see, it will accompany me into my second childhood. It has provided me with an absorbing hobby which made museums the joy of my boyhood and regimental messes, with their untold treasures, that of my days in the old Indian Army. Like a magpie I have collected militaria upon no particular plan—postcards, buttons, weapons, castle guides and so on. For one reason or another, much of my collection is now gone, but at one time the accumulation of hitherto somewhat useless knowledge became of practical use, when I ran the famous *Tradition* military antique shop in Dover Street, London, for several years. I then edited the late lamented *Tradition* magazine until, after ten years, inflation killed it. Some minor literary works, including a *History of the Gurkhas* and *Costume of the British Army in the Crimea,* filled in the time not occupied by historical re-enactment of the battles of the English Civil War with the Sealed Knot, of which I was a Director for some ten years.

In 1917, when my father came home on leave from France, he bought me a box of the famous Britain's toy soldiers—Lifeguards in the panoply of pre-1914 full dress. We went to the seaside, and there I collected cigarette cards from the convalescing soldiers who, in their bright blue hospital suits and red ties, sat enjoying the sun. But I collected more than cigarette cards: my constitution had been unwittingly undermined by the Lifeguards, and I fell victim to a virulent form of infection—*collectomania militaria omnivera*—from which I have suffered ever since. Acute financial debility has at times permitted some regression, but with any increase in funds the disease would, like recurrent malaria, break out again, with the inevitable aftermath of acute spasms of pain in the purse. It was many years before I was able to comprehend the proposition that so vital and momentous a transaction as the purchase of a red coat or a sword should be postponed, let alone abandoned, in favour of settling some contemptible triviality such as a gas bill. Eventually, like St Paul on the road to Damascus, I was struck by the light of a Divine Revelation: you don't have to possess things in order to enjoy them. In a world of escalating prices this is a comforting philosophy; it enables me to concentrate on resisting, with only moderate success, the continuous barrage of temptation put out by the publishing houses (present company not excepted).

The life of a man who collects things has extra dimensions, one more for each of the things he collects, although most people tend to specialise for reasons of economy if no other. I am all for specialisation within reasonable limits, since specialists are apt to discover and, it is to be hoped, record information helpful to others. There is, however, one form of specialisation which I abhor, which I class as vandalism pure and simple, and of which I will quote examples that have occurred in my presence. I have seen the buckle cut away from a fine gold-laced sword belt, which was then discarded, because the purchaser collected only buckles and badges; I have seen buttons and collar and shoulder badges ripped from a jacket in near-mint condition for the same reason; and finally, I have watched while the lining was torn from a fine cavalry officer's pre-1914 full dress helmet, an act which remains incomprehensible to this day.

When museums all over the world are trying to assemble complete specimens, it seems a pity that collectors, whom one would expect to treasure such things, should destroy them. Strangely, it is often the commonest or least valued item that becomes the rarest. Several years ago, the National Army Museum in London mounted a splendid special exhibition on the former Indian cavalry. It consisted of a number of lifesize reconstructions, one of which showed a group of officers in a mess anteroom. I asked the curator where the mess orderly was, with the drinks. He replied that the museum had no servant's uniform. On my return home, the *Raj* went into reverse as I starched and ironed my former bearer's dress uniform. I presented it to the museum, and the mess scene sprang to life as drinks were handed round. (Another irritating type of specialist, you may be thinking, is the one who cannot resist an ego-boosting one-upmanship trip!)

It is the fashion in certain quarters nowadays to decry all interest in things military, as though collecting badges, buttons or firearms will bring forth a race of homicidal maniacs. I disagree. Admittedly man (and, of course, woman) has an inordinate degree of aggression, which is what makes him (and, of course, her) the most successful predator this planet has ever seen. But you will not find much evidence of violent aggression among collectors of militaria. Is there, perhaps, some lesson to be learned from the fact that dedicated practitioners of the martial arts are usually the gentlest of men, not given to taking life for kicks as in hunting, shooting and fishing?

Having observed the field of collecting for more than half

a century, it seems to me that life is now rather more difficult for the collector that it was, and this is largely because there are apparently far more of them, attempting to acquire only marginally more items at ever-increasing costs. Certain kinds of militaria have always been comparatively expensive; for example, armour, which fell out of favour in the 17th Century, lasting a few more years in portraits, but came back as a top-flight collectable during the Gothic Revival of the 19th Century. As I write, armour has taken off into the outer spaces of finance with the sale of a suit at Sotheby's in London for £2,000,000. Gone now are the days when two or three dozen military buttons on a piece of string could be had for less than the price of five cigarettes, a fine sword and scabbard for less than the price of 100 cigarettes, and a Navy Colt in a velvet-lined mahogany case with all the accessories for less than the cost of 600 cigarettes. These were all purchases I made when a schoolboy—and the price of cigarettes used as a comparison is the price they were *then!*

When I went to India in the 1930s to indulge in the last days of horsed cavalry, I had high hopes, but I found the bazaars of India, with its great military traditions, surprisingly barren. The climate, the moths, the white ants are lethal so far as perishable materials are concerned. I remember my joy, rapidly replaced with envy, at the sight of a Sikh bus driver coming over the Afghan border at the head of the Khyber Pass wearing a British Guards' officer's beribboned frock coat—with a twelve-inch gap at the waistline. The prospect for non-perishable items was little better, for metal was used and re-used. One day, in Bombay, I called upon a jeweller at the very moment he put a set of fine silver medals with bars, which had belonged to an old Indian officer, into the melting pan. I waited for them to cool off and left in triumph with my purchase.

In Europe, vast areas were devastated by war and others have been swept away by post-war redevelopment. Gone are the junk shops which were part of the collector's world. A classic example is the English city of Birmingham, a collector's paradise between the World Wars. I remember one little house there in which every room was lined with shelves, on which were boxes filled with all the brass and steel fittings for Tower muskets, the famous "Brown Bess" musket used by the British Army during and for some years after the Napoleonic Wars. They had been inherited from the owner's great-great-grandfather, who had supplied them to the governments of the day.

There are still bargains to be found but, excluding beginners' luck, for which I have the greatest respect, and all other things being equal, the chances of picking up a bargain or even making a purchase at a reasonable price depend largely on knowledge of one's field—to which I would add the intuition which undoubtedly develops with practice. Reading up one's subject is one of the great pleasures of collecting and, in spite of rising prices for our collectables, we are fortunate in that publishers, in their efforts to keep abreast of the demand from the ever-increasing band of collectors, have produced many reference works, good, bad and indifferent.

This book is intended as an introduction to militaria collecting, and there is no need to apologise for the production of a primer when it is of the quality of this volume. I feel it only fair at this stage to declare an interest: I was in at the initial stages of planning this book, but because of serious illness I had to drop out. I regret this even more now that I see the final work. However, I draw some satisfaction from seeing that the enlightened publishers agreed to my suggestion that coverage should reach back to the beginning of the 20th Century, to the last days of a now-vanished world. Inevitably in a work of this type, so much has to be left out, and I wonder how many readers will appreciate with what heart-searching a decision to exclude some item is taken? In this case the publishers have wisely decided to concentrate on items that the *average* collector in the Western world can hope to find and acquire. What I find excellent is the high order of the layout, the clear illustrations and pithy text. The percipient reader who has stayed with me so far will by now have begun to suspect that I like this book. He would be right, and so I commend it to you.

Lieutenant-Colonel (retired) J.B.R. Nicholson
Ivychurch, Romney Marsh, Kent May 1983

Introduction

Joe Lyndhurst, Warnham War Museum

There is no doubt that I was very flattered to be asked to write a general introduction to this magnificent publication on collecting modern militaria. Collector I may be, but I have never considered myself qualified to write on the subject, and when I was asked to do so I felt the waters rising over my head. I took a deep breath and struck out for the shore. I am, after all, the owner of a military museum, so I reasoned I must know something about the subject, and with this in mind I began.

Collecting things generally is very much in the tradition and the national character of Britain, where it is not considered strange or eccentric to collect any mortal thing. The same is true of the United States, I believe; and although the magpie instinct is, perhaps, not so great among the nations of Western Europe, there are certainly some fine public and private collections in these countries.

The small private collections which have formed the nucleus of many of the fine British museums of today were created by people of foresight; not always of great means, but with respect for the past and hope for the future. Britain owes a great deal to men like R.G.J. Nash, whose collection of World War I aircraft is now distributed between at least three of the best British aviation museums, and Richard Ormond Shuttleworth, aviator and collector, whose collection was formed by Lady Shuttleworth into the renowned Shuttleworth Trust, based at Old Warden in Bedfordshire, in order to perpetuate his memory.

LIVING IN THE PAST

Museums in their modern form, as places of both instruction and entertainment for the general public, largely evolved in the 19th Century, reflecting that era's great pride in the achievements of the Industrial Revolution and, in Britain, the great national pride in the dominant position of Victorian England in the world. Today, museums are opening at a rate previously unknown. In the even greater industrial revolution of the 20th Century, with the rapid development of almost every commodity known to man, artifacts become obsolete and are disposed of quickly. But in an effort to hold back the clock, or at least to slow it down, people are tending to turn to the older, familiar things in a wave of nostalgia and, for relaxation, they resort to collecting items of yesteryear. Living in the past has become a therapy.

My earliest recollection of collecting dates from the 1930s, and is of 1/72-scale military aircraft in kit form by Sky Birds, a company that produced a whole range of aircraft constructed from hard wood and brass, from World War I to early World War II types such as the prototype Spitfire and Blenheim. At a very early age I was as familiar with the range, performance and fuel load of a Westland Wallace as the next man. The next milestone along the path was the pre-War air displays held on Empire Day, where I was introduced to the real thing: Hawker Furies firing their twin Vickers guns at the butts, over the roar of Rolls-Royce Kestrel engines. I have some of those expended cases and links to this day. From that time on, I was firmly hooked on collecting modern militaria.

MILITARY VEHICLES

The first military vehicle to take my fancy, in about 1942, was a Canadian Army jeep, a robust little vehicle of such style that, I think, it did more than anything to turn my attention to collecting the military vehicles that now form the major part of the Warnham War Museum. Collectors of militaria are not necessarily former military men, nor, contrary to what amateur psychology may suggest, need they be working out some inherent aggressive trait. Most are, like myself, people who admire the achievements of some particular individual or group or the spirit of some period of history, and

Far left: *Serious collectors of militaria are unlikely to forget that their pastime has its base in grim reality. Here, US Infantrymen examine the bodies of German Waffen-SS troops killed in a fire-fight in the Dachau area of Germany, April 1945.*

Left: *Collectors' interest will centre on the protective headgear in this photograph of a US Army tank crew, 1944; examples of the tanker's helmet worn by the commander and driver and the steel helmet of the third crewman are shown in colour on* pages 104-105.

Above: *British cap badges offer a very rich field for the collector. In this World War I group, the men seated at the ends of the row wear the badge of the Army Service Corps; the rest, the rarer badge of the Army Cyclist Corps, in existence only from 1914 to 1919.*

Right: *An Officer of the Royal Horse Artillery in full dress, c1912; note the Royal Artillery pattern sword with dress knot and the plumed headgear. Within a few years, full dress would give way to service dress of the kind seen in the photograph to the left on this page.*

therefore wish to build up a monument to the particular group or period. In my case, the Canadian jeep was the trigger that set me off on the all-absorbing pastime of vehicle collecting.

RESTORING AND PRESERVING

By the late 1950s military vehicles of World War II were beginning to be in short supply. The first ex-service vehicles of all types sold as surplus after the War were eagerly bought up and converted as far as possible for use in civilian life, but owing to their comparatively small load capacity and low gearing they were in the main scrapped and replaced as soon as possible by the new commercial vehicles that were beginning to come on to the market. The exceptions were vehicles with good off-the-road performance; these enjoyed a long life as recovery vehicles for garages and agricultural contractors, as snowploughs for County Councils and, in the case of the Army's tank recovery vehicles, for the use of showmen in moving fairground equipment.

There were exceptions. A number of the military vehicles manufactured during the 1940s are still in use throughout the world: notably the "jeep taxis" in the Philippines, the International and White half-tracks still in Israeli service, and the GMC 6x6s in Red China. The Scandinavian countries and France still operate a number of US-built vehicles of the 1940s; the same vehicles are increasingly rare in the UK. However, I know of a number still in regular use in the UK. My own jeeps and other vehicles are kept road-worthy and are in great demand for shows of all kinds—and one of the contributors to this book keeps a US Army White half-track parked outside his home in the London suburbs. It is safe to say that anyone who has used vehicles of this vintage regards their passing with something more than nostalgia.

RE-ENACTMENT SOCIETIES

Some collectors, like myself, have dedicated themselves to restoring and preserving military vehicles. Soft-skinned vehicles of World War II are still to be found in scrapyards in the UK, but they are usually pretty far gone; the collector will be fortunate to find one in salvageable condition these days, even if he widens his search to the continent of Europe. I suppose that, strictly speaking, the collection and restoration of vehicles is outside the scope of this book, but I have introduced the subject to illustrate just why a number of people have started to collect militaria; for some it is a hobby, for others something more.

It is worth mentioning here the various military re-enactment societies, some of which have come into being largely because of the activities of vehicle restorers. These enthusiasts, having gone to great lengths not only in restoring their vehicles but in finding the correct equipment for them and for the soldiers of World War II—not only uniforms and weapons, but also such equipment as cooking stoves, rations and tools—often wish to use their vehicles in conditions which simulate the "real thing" as closely as possible. Thus, re-enactment societies, which are not limited to World War II but range from Roman times up to date, do battle on opposing sides during the summer months. Whether the encounters portrayed are those of the American Civil War, the English Civil War (as in the case of the well known "Sealed Knot" society of the UK), or World War II, they are of great interest to the collector and student of militaria, because of the insight one obtains from actually experiencing the appropriate equipment in use.

STARTING A COLLECTION

There are several things that the newcomer to the field must consider before starting a collection of militaria; notably, the space available for storage, the amount of cash available, and the period of specialisation. As I have

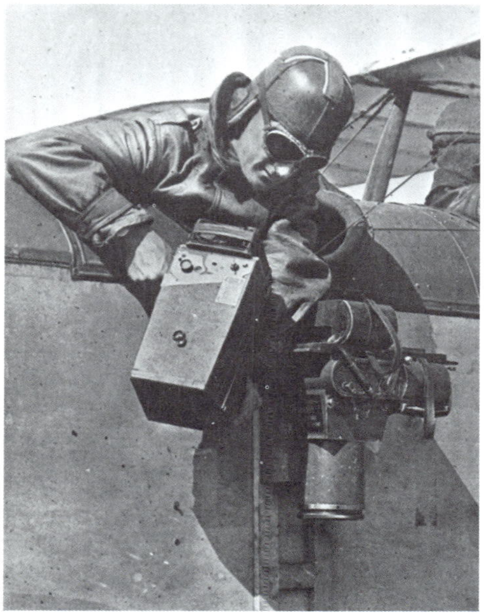

Above: *US Army aviators demonstrate the techniques of aerial photography in 1919. Note the headgear of "crash-helmet" type; a selection of flying helmets is shown on* pages 106-107.

Left: *In service dress, then fairly recently introduced (note cloth shoulder titles), men of the Royal Scots Fusiliers pose with a heliograph, c1912. Signals equipment is shown on* pages 202-203.

Right: *A wide range of military headgear is "modelled" by Officers of the multi-national China Relief Expedition, 1900-1903; note the French kepi and the side cap of the British representative.*

already suggested, it is usually a fascination with one particular period of history that starts the collector off. Here, we must assume that the modern period, 1900-1983, from the Boer War to Vietnam and the Falklands, is the one chosen.

If both space and cash are limited, the new collector would do well to think of collecting cloth insignia (see *pages 108-139*). These are generally still not too hard to find, are comparatively cheap, and make a colourful display, either in frames or in an album. The different types of unit patches worn by the Allies alone during World War II must run into the thousands, and some are now getting difficult to find, particularly the early printed-on-cloth patterns, so there is the chance of acquiring rarities. Cloth insignia of the flying and naval services, which are numerically smaller than the land forces, have great appeal, particularly those of the United States Air Forces (see *pages 124-125*), and these are still surprisingly easily available.

With space and general availability still in mind, buttons (see *pages 46-55*) would probably come next on the list for the collector with limited room to display or store specimens. Since this has been a popular collecting field for many years, there is a wealth of reference material from which to seek guidance. (Some sources are listed in the "Bibliography" on *pages 206-207.*) Many old soldiers retained the buttons and badges of their former regiments even if they preserved no other mementoes, thus assuring a supply for the collector; and extensive button collections regularly come under the hammer at auctions devoted solely to military buttons; it is a well-organised field and probably one of the easiest for the new collector to start with.

PRICE RANGE

At the top end of the market, but still bearing in mind the space problem, come campaign and gallantry medals. Here, prices range from the cost of a packet of cigarettes to the value of a luxury yacht; determined, of course, by the scarcity of the medal in question. Fine medals are without doubt the costliest items within the scope of this book; outside that scope, the highest prices are probably those paid for aircraft of World War II—in the UK, more than £200,000 ($304,000) has been paid for a Spitfire, an aircraft that will forever capture the imagination of military enthusiasts and which may be considered the ultimate in piston-engined fighters.

On the subject of expense, it should be noted that although our contributors have in many cases indicated items that are likely to be costly and hard to find, as well as those that are easily found and relatively inexpensive, no detailed price guide has been attempted in this book. It is obvious that prices will vary from time to time and from country to country, and it was felt that a scale of values based on current UK prices would be of little use to collectors elsewhere or, indeed, to collectors in the UK within a year or so of publication.

Many militaria dealers in the UK, Europe and the United States issue regular price lists—in the case of the major auction houses, catalogues and price lists often take the form of well-illustrated books—and these should be generally available to collectors. Remember, however, that in many cases values will be subjective: one collector's "rubbish" is another collector's "treasure" and, generally-acknowledged items of value apart, many pieces of militaria are worth just as much, or as little, as a collector is prepared to pay.

CAP BADGES

To return to items within reach of the average collector: one of the most popular, and perhaps one of the longest-established collecting fields is that of cap badges, an activity which probably began in the UK during the twilight of the Victorian era, with the passing of the helmet plate and the coming of

the field service uniform and cap, which required a smaller badge, easily fitted by means of a tongue or bar, for the field service cap, glengarry, side cap or, later, beret.

The range of British cap badges is immense. A collector may choose to collect the badges of the Army of World War I, including the Commonwealth and Empire troops and the new units formed in 1914-1918 war—such as the Machine Gun Corps, the Cyclist battalions and the Royal Flying Corps. An excellent cross-section of British cap badges of the 20th Century is shown on *pages 46-55,* where the many variations to be found—the change of monarchs that changed the designs, the shortage of strategic metals that altered the material, the disbandment or amalgamation of famous regiments—will be apparent. The collector, of course, delights in such variations, and will always hope to find one which has escaped the attention of the compilers of the excellent reference books on this subject.

A good, and inexpensive, way to begin the collection of cap badges is to check up on one's relatives: in almost every family there is someone who has served in the armed forces at some time, and ex-servicemen almost always keep a cap badge or two. Bob Cayley, whose fine collection of British cap badges, buttons and shoulder titles is featured in this book, is himself an old soldier and began his collection with his father's World War I insignia and his own from World War II.

PROGRESS TO HEADGEAR

During the search through flea markets, militaria fairs and auctions for British cap badges, the collector will no doubt encounter the metal insignia of other countries, with the inevitable result that further collections will be started—it is only the most resolute and single-minded who never deviate from the original theme. Few collectors will be able to resist such attractive pieces as the large and impressive US badges shown on *pages 60-61* and the brightly-enamelled, rather exotic unit insignia of the French Foreign Legion (*pages 62-65*), or the somewhat sinister lure of the badges and combat awards of the Third Reich (*pages 66-67*). Quite different, but no less fascinating, are the "Sweetheart Badges" shown on *pages 56-59:* here, cap badges and other insignia have been mounted as brooches, or commercially reproduced in miniature, for servicemen to give as gifts to their loved ones. These are particularly popular with lady collectors: the examples shown come from the collection of my wife, Yvonne Lyndhurst, and from that of Mrs Vannessa Walker, wife of a major contributor to our coverage of cloth insignia (*pages 108-139*).

There is also the "danger" (to the collector's available space, as well as to his pocket) that the collection will soon progress beyond badges alone. It is a fairly safe bet that the badge collector will wish to purchase a cap or two, with the object of displaying some particularly treasured badge, and will thus be led into yet another branch of militaria. The collection of military headgear of the 20th Century opens the floodgates to dress and service headgear, combat and protective wear, ranging from the exotic helmets of Imperial Germany (*pages 98-99*) to flying helmets (*pages 106-107*), the one steeped in Germanic tradition, the other strictly functional.

CLOTH AND STEEL

The appeal of such items of dress headgear as the richly-coloured British peaked caps and side caps shown on *pages 86-89* (again from the collection of a lady, Miss Margaret E. Nobbs) is obvious, and the US forces' dress, service and garrison caps shown on *pages 92-95* are hardly less attractive. German dress and service headgear of the Imperial and Third Reich periods (*pages 90-101*) is likely to be more difficult to find and more expensive,

Left: *Christmas in Korea, 1950; note particularly the GI mess-tins and the Soviet-style Cold Weather Caps (see 1, page 92) of these American soldiers.*

Far right: *A serviceman's nightmare—but the wealth of equipment is a collector's dream: kit inspection for a long-serving soldier (note inverted chevrons and medal ribbons), Oxfordshire and Buckinghamshire Light Infantry, c1920.*

Right: *Experimental British service dress of 1937, modelled by a Company Sergeant-Major of the Welch Regiment.*

Below: *With Springfield rifles piled, men of the American Expeditionary Force parade in the UK. 1917, note characteristic Campaign Hats and US Army collar discs.*

but berets (*pages 90-91*) now worn by services of most nations, may be fairly easily obtained. No representative collection of military headgear would be complete without at least one example of the famous French kepi (*pages 96-97*), which has been in wear since the first half of the 19th Century.

Nor should the collector overlook the possibility of collecting steel helmets (*pages 102-105*). To the uninitiated, it may appear that there is little difference between one steel helmet and another, but a study of our photographs and captions will correct this view. By no means a new idea, the steel helmet was re-adopted by the German and British Armies in 1916, for the trench warfare of World War I, and used with some modifications throughout World War II. The United States, after initially adopting the British helmet, produced its own distinctive design in the early part of World War II. Originally intended for the Marine Corps, it proved so successful, with its detachable liner that allowed the helmet to be used to wash in or carry water, that it was adopted by all United States services and is still in use in that country and in others throughout the world.

Finally, on the subject of headgear collecting, gas-masks (*pages 190-191*) often appeal to young collectors in the UK, principally because of the abundant supply of British masks and the many types of civilian examples. Once having embarked on this speciality, the collector will almost inevitably start hunting down foreign types, and will extend his researches to the early examples produced in World War I, when poison gas was used with such devastating effect. Gas masks designed for service animals were produced at this period; they are now extremely rare.

PRINTED MATERIAL

As our selection on *pages 162-179* shows, militaria in the form of words on paper comes in a bewildering variety of types, ranging from service manuals to autograph letters and from newspapers to maps. Service training manuals are a "must" for the collector of militaria, as much for the information one can glean from them as for their intrinsic "collectability", but they are, unfortunately, increasingly difficult to find. Military manuals were not, of course, intended for very wide circulation—and paper of all sorts went for salvage in wartime; not excepting the poster that proclaimed "Waste Paper Ammunition of War"! Here, I can only mention some aspects of this field that have always held a special appeal for me.

Entertainment was of prime importance during wartime, and concerts and camp shows were always encouraged among troops of all nations. Mobile cinemas and local concert parties did much to relieve the boredom of camp life and lift morale. So important was this considered that both the USA (with USO) and Britain, with ENSA (Entertainment of the National Services Association), formed official entertainment organisations. ENSA was under the direction of the impresario Basil Dean, with headquarters in the Theatre Royal, Drury Lane, London. To collect the programmes of ENSA's shows is fascinating, for they were staged throughout the world, wherever Allied troops were stationed. Performers and technicians serving near the battle fronts wore service dress similar to that of war correspondents; the collar insignia, or "collar dogs", showed the acronym ENSA—which some comedian of the day christened "Every Night Something Awful"!

Under the general heading of entertainment and morale boosting material also falls the sheet music of the time, as well as magazines and newspapers. Our selection in the latter category ranges from a German humorous magazine of World War I (*pages 178-179*), to the newspapers specially produced for British and American servicemen during World War II (*pages 170-175*). These are becoming increasingly hard to find, as are the propaganda leaflets

produced by the million in wartime for air-dropping over enemy territory. A further interesting specialisation in the field of printed materials would be postal items, including Christmas cards, that showed some ingenuity of home-made improvisation during wartime, and the many methods employed by the forces' postal services to keep loved ones in touch. Military philately, a highly-specialised field, was regretfully decided to be beyond the compass of this book.

SERVICEMEN'S GIFTS

Most servicemen have always attempted to bring home souvenirs and gifts; probably the best-known and now most valuable items of this kind are the World War I items made from scrap metal and now generally referred to as "Trench Art". Examples from both World Wars and from conflicts ranging from Imperial Germany's colonial campaigns to the Falklands' battles of 1982 are shown on *pages 182-185*.

Prisoners-of-war often passed the wearisome time of captivity by practising handicrafts: one of the most prized items in my personal collection is a tiny model of a corvette made out of bone by a French prisoner during the Napoleonic Wars. This, of course, is outside our period, but I also have an amused regard for such PoW "art" as the improper caricature of Adolf Hitler shown at (35) on *page 183*. This was made by an Italian PoW in England during World War II: among these men, a thriving cottage industry existed, as they produced gifts that could be exchanged for small luxuries.

Of great delicacy and charm are the embroidered representations, some on postcards and handkerchiefs (*pages 186-189*), others worked on large canvases in silk or wool, of regimental badges and other unit or national insignia. The postcards were sent by servicemen as gifts and tokens of affection. Collectable for their decorative value alone, they have increased in monetary value in recent years.

PERSONAL EFFECTS

The personal kit of the fighting man always has a story to tell. One marvels at the small amount of personal possessions an infantryman had. Of course, his gear was kept to a minimum because he had to carry it wherever he went—barracks or foxhole, desert or city—and his shaving brush and razor, soap, towel, spare socks and needle and thread made up almost the whole of his non-combat equipment. All these items are shown on pages *194-195* in our "Miscellany" section. Among the serviceman's other personal possessions, one would expect to find his pay-book, the personal record a soldier always keeps with him, a letter from home, a photograph or two, a cigarette tin (or maybe a handsome case made out of scrap aluminium from a downed aircraft; like the Japanese soldier's handiwork shown at (29) on *pages 182-183*) and, of course a cigarette lighter that could be lowered into the platoon truck's petrol tank when it ran dry. Last, but by no means least, is the identity disc or discs worn around the neck by every serviceman, engraved with name, number, blood group and religion: information to be used in case of death or injury. These were always referred to as "dog tags"—because that is precisely what they looked like.

Sailors fared a little better, although they lived in cramped accommodation aboard ship, and usually had more personal kit stored in a "ditty box". Airmen on large, permanent bases did best of all, with what were four-star quarters compared to those of other services and, in consequence, a deal more personal kit. But soldier, sailor and airman alike all shared the "dog tags", the pay book, the button sticks, the shaving brush—and the perpetual lack of privacy.

It has been said that an army marches on its stomach, and there are few things that servicemen of all nationalities have groused about more than food.

But during World War II, food for the services was for the first time given proper consideration by nutritional experts, in an attempt to ensure that both "cookhouse" meals and combat rations would give the best possible nutritional value. Much research and innovation by food processors resulted in great advances in packaging and preserving. One object was to save shipping space, in view of the enormous quantities needed in the war zones and, towards the end of hostilities, by the near-starving peoples of Occupied Europe. A cross-section of rations, both military and civilian, is shown on *pages 196-197*.

THE "HOME FRONT"

It is worthwhile remembering that during World War II, 1939-45, in spite of grave shortages of food and severe rationing restrictions, the health of the British nation improved. On the "Home Front", a field in which collectors are increasingly interested, thanks to the "nostalgia boom" I have already mentioned, there appeared innumerable magazine and newspaper articles and cook books aimed at stretching the rations by using alternative ingredients and growing your own food. Growing mushrooms under the stairs, for example, replanting your flowerbeds with vegetables and fruit, keeping a few hens for eggs—the UK ration was four eggs a month in 1943!—or rabbit-rearing for those fortunate enough to live in a semi-rural area where they could gather green fodder. There were leaflets, too, extolling the value of dandelions and nettles as green vegetables and listing edible nuts, berries and fungi, thus generally bringing long-lost skills back to the British kitchen. A "wartime cookery" collection could be well worthwhile: one might even try out the recipes!

MEDICS AND MEDICINES

Wartime medical preparations (*pages 198-199*), when viewed from the 1980s, seem rather elementary, but we should remember that many of the enormous advances made in surgery and medicine, including skin grafting and the control of gangrene with the emergence of penicillin, were in no small measure the result of intensive work during World War II. This makes medical items of great interest to the collector, whether they are sought in their own right or to complement a collection of nursing insignia, Red Cross items, and uniforms, such as those of the nursing services and the Royal Army Medical Corps (RAMC). The First Aid kits carried on aircraft, for example, were very comprehensive, for a crew member might have to deal with serious wounds and burns over the target area, many hours away from base. These kits would sometimes make the difference between life and death.

The tools used by the fighting man constitute another area of interest to the collector. As may be seen from the selection on pages *204-205,* the wire cutters and entrenching tools of various nations differ only slightly in principle but quite widely in design and manufacture. Entrenching tools, a necessity for the infantryman since the mid-19th Century, one would suppose to be much of a muchness; in fact, they incorporate all kinds of attachments to make them as versatile as possible, presenting great variation to the attention of the collector in this field.

WEAPONS OF WAR

The ultimate tool of the ordinary soldier is, of course, his rifle or hand gun. Firearms have always fascinated collectors and non-collectors alike, and the workmanship, the effectiveness and even the aesthetic qualities of weapons are subjects for discussion whenever collectors meet. Although the collection of firearms is hedged around with legal restrictions (summarised on pages *140-142*) which may deter would-be collectors, this book has a fine section on the subject of firearms and ammunition (*pages 140-*

Above: *Two Australian soldiers of the United Nations forces in Korea, 1951, support a US soldier weakened after captivity in Communist hands. Note spare magazines and grenades (see pages 158-161) on the webbing belts.*

Above left: *New Zealand sniper with a telescopic sight on a No 4 rifle (see pages 150-151), in action at Cassino, Italy, March 1944. Sniper versions of the Lee-Enfield are still fairly easily found by firearms' collectors.*

Left: *US Marines in street-fighting at Naha, Okinawa, May 1945. Note the rifle-grenades on the pack of the central man, and the attachment for firing them at the muzzle of his Garand M1 carbine.*

Above: *A US Marine gives a wounded comrade a drink from his canteen; Peleliu Island, Pacific, 1944. Note steel helmet (see pages 104-105) with camouflage cover and, strapped to pack, the entrenching tool (see pages 204-205), an essential part of the combat soldier's equipment since the mid-19th Century.*

Above left: *US Army Medical Orderly (note Red Cross armband and, above it, cloth patch of the 3rd Infantry Division) gives a blood transfusion; Sicily, 1943. Like the "Medic", the wounded man will have around his neck "dog tags" engraved with personal details, to be referred to in just such an emergency.*

161), and I think it best to let the photographs speak for themselves.

Coming close to firearms as a field of special interest, and rather easier to collect, are edged weapons, both utilitarian and ceremonial. The splendid selection shown on *pages 16-41* ranges from ceremonial swords with finely-engraved blades to jack-knives and present-day survival knives. Perhaps the simplest and cheapest edged weapons to collect are bayonets—and these are by no means the least interesting of the weapons shown in this book. The principle of the bayonet reaches back to the 17th Century, when infantrymen tied or wedged in knives to the muzzles of matchlocks, undoubtedly an improvisation born of desperation. The knife bayonet as we know it today is considered to have evolved in the Americas in the 1860s, signalling the demise of the awkward socket bayonet. It is extraordinary that the bayonet still survives in service: it has generally proved to be a poor instrument for hand-to-hand fighting (although some bayonets have been designed for use as fighting knives when removed from the rifle) and its use with a modern self-loading rifle would seem to be superfluous. However, an enormous variety of bayonets is available to the collector: they have been in combat use for some 300 years and are likely to continue at least in ceremonial use in years to come.

I am sure that few readers, having studied this book, will not at least consider joining the fraternity of militaria collectors. Although I have suggested that a new collector should be guided into a speciality by what he can afford, as well as by what most interests him (or her), I cannot honestly say that I know of a collector who has turned down anything he really wanted on grounds of expense alone.

A PIECE OF HISTORY

I know that I have spent too much, too many times, and I suspect that most collectors have also spent on their hobby money that should have gone for some necessity. But the satisfaction that my collection has given me is inestimable. I have been able to establish my own museum, which has been graced by the visit of Air Chief Marshal Sir Arthur "Bomber" Harris, Air Officer Commanding in Chief, Bomber Command, RAF, during World War II, and through my collecting activities I have had the honour of meeting HRH the Duke of Edinburgh.

My advice to the collector can only be to follow your heart and inclination, collect whatever appeals to you, learn all that you can about it and treasure it; it is yours—and it is a piece of history.

Edged Weapons

Gordon Gardiner

"The history of the sword is the history of humanity" wrote Sir Richard Burton (1821-1890), the famous traveller, author and eccentric, himself a master swordsman, in *The Book of the Sword* (1884); "to surrender the sword is submission", he continued, ". . . to break the sword is degradation. To kiss the sword . . . is the highest form of oath and homage."

Since it first reached its recognisable form in the Bronze Age, the sword has been the "Queen of Weapons", and even with the advent of black powder and the intrusion on the battlefield by cannon and musket, the sword continued to hold sway with the cavalry of the world. Standardisation of style and manufacture was introduced for the military swords of European and North American forces in the later 18th Century, and although the development of the metallic cartridge and multi-shot firearms in the 19th Century diminished the role of the edged weapon, World War I, with its artillery barrages of unparalleled ferocity and its massed machine guns, still saw the sword as a standard weapon for the cavalry and, at the beginning of war at any rate, for infantry officers also.

As late as 1942, during World War II, some British Yeomanry regiments, now mounted in tanks rather than on horses, had racks fitted in their armoured vehicles for their Troopers' Swords of 1908 Pattern! In Japan, the sword occupies a central position in the code of *Bushido* ("the way of warrior") —"The sword is the soul of the *Samurai*" is a famous aphorism—and during World War II officers of the Imperial Army and Navy invariably carried their swords (some of them ancestral blades of great age, in modern mounts; others Arsenal (factory-made) blades) in action and used them whenever possible: on Okinawa, a US tank commander reported an attack on his AFV by five sword-wielding Japanese officers; off Guadalcanal, the men of two New Zealand corvettes engaging a surfaced Japanese submarine found themselves repulsing a sword-wielding boarding party. And although in the West the sword is now regarded as being purely a ceremonial weapon, whose style and patterns have changed little since the beginning of the century, there appeared in the British Press a few years ago a dramatic photograph of a Trooper of the Life Guards holding back demonstrators in Whitehall, London, with the threat of his drawn sword.

The sword is, of course, not the only edged weapon known to warfare. There are others of a more modern nature: the bayonet, the various types of fighting knives, and specialised knives such as those issued to aircrews, survival knives, and the like. All these have been products of necessity. Many have claimed that the bayonet and the knife are outdated; few who have had recourse to them, even in the recent Falklands conflict, would agree. The basic blade will survive, despite "the bomb". It is one of man's oldest tools, and surely one of the most versatile.

SWORDS AND DAGGERS

In the period under review (1900-1983), the sword was, in British service, a ceremonial weapon, with its edge left unsharpened in peacetime; but on mobilisation in 1914 and 1939, British officers sent batmen with their swords to the regimental armourer, where the cutting edge was ground on. Other Ranks' bayonets were treated in the same way, as were other edged weapons such as cutlasses, lances (in World War I) and boarding pikes. Pipers of Scottish regiments sent in their dirks, and Pioneers (prior to 1903) their traditional saw-backed swords.

Dirks or daggers have always been a traditional item of Scottish Highland wear as well as being the arm of Mid-

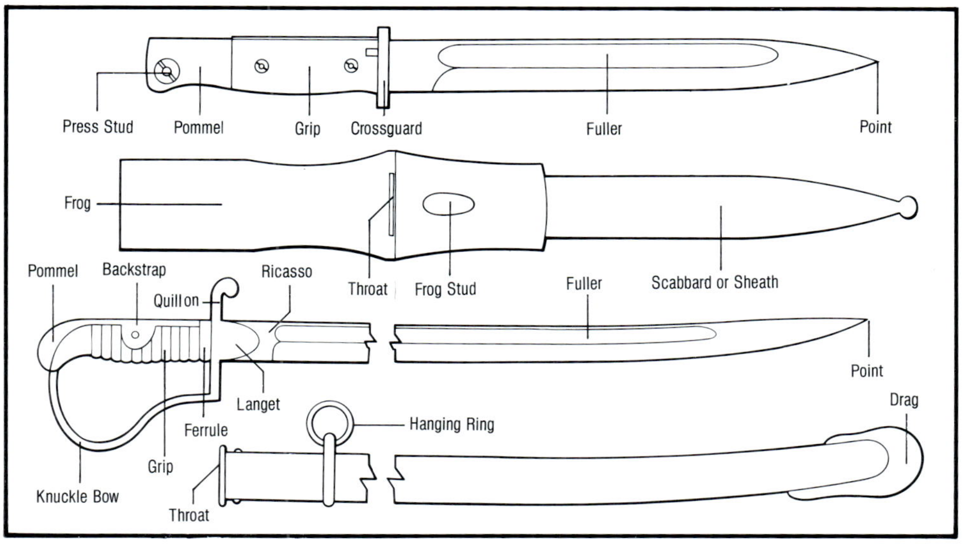

Above: *Although seemingly superfluous, given the firepower of modern arms, the bayonet remains in service. The Italian Army's Alpine Troops parade with fixed bayonets on BM 59 Mark Ital TA rifles.*

Left: *When the men of D Squadron, 19th Hussars (Queen Alexandra's Own), posed for the photographer in c1904, with Cavalry Troopers' Swords, 1899 Pattern, the sword was still considered to be a practical combat weapon; so much so that in 1906 a special War Office Committee was set up to conduct the discussions and tests that led to the introduction of the 1908 Pattern Sword.*

Right: *The sword in its modern role, carried by Officers for ceremonial use as Princess Alexandra takes the Sovereign's Parade at the Royal Military Academy, Sandhurst, in July 1968.*

shipmen in the Royal Navy; a number of other navies adopted the dirk for officers below flag rank, and many still wear it. In Nazi Germany during the 1930s the proliferation of dress uniforms for both political and military organisations gave rise to many new designs for dirks and daggers. Among collectors, the swords and especially the numerous daggers of the Third Reich exert a tremendous fascination. Many Western sword blades are etched with such designs as scrolls, cyphers, unit badges, monograms and the like, but German manufacturers especially offered a great varity of etched designs, and fine presentation swords are more likely to be found for Germany than for other nations. Third Reich dress daggers and sidearms will be found to vary widely in quality: some organisations insisted on strict quality control for their daggers, which will have the blades stamped with the mark "RZM" (*Reichszeugmeisterei;* "Reich Master of Ordnance"). Because of the great interest shown in these weapons by collectors all over the world, there are now factories, in Spain and elsewhere, turning out reproductions which may well deceive the new collector who has not carefully researched the subject. Another, less collected, area is the daggers of the Italian Fascist period, from the 1920s until World War II, which reflect the same ideology as their German counterparts—although the Italians seem never to have had, in this century, the flare for the military style that Germany, with its main production centre at Solingen, possessed.

KNIVES AND BAYONETS

Fighting knives were largely a product of World War I, when the trench warfare on the Western Front gave rise to a need for silent and deadly weapons. At first, a large array of hunting knives, cut-down bayonets and home-made daggers was produced unofficially, but by the end of the War, the Americans and French at least had produced official patterns, the most notable being the US 1917 and 1918 Pattern knives, with brass knuckle-duster grips. During World War II, with its Commandos and other "special forces", the art of knife-fighting became more scientific and special knives were designed. The most famous was the British Fairbairn-Sykes ("FS" or "Commando Knife") fighting knife, made from 1941 onward and designed by the two officers whose names it bears in conjunction with the famous swordmaker Wilkinson Sword Company. At the same period, Germany produced its unique gravity knife, particularly favoured by paratroopers.

The largest category of collectable edged weapons for the period under review must be that of the bayonet which, unlike the sword, continued as a functional weapon in the 20th Century and thus saw many developments in type. During World War I, the bayonet took on two roles: its prime purpose, as a weapon, to covert the rifle into a pike (the spirit of the pike died hard!); and as a handy weapon for opening ration boxes, cutting firewood and similar tasks. The tendency was for the blade to become shorter, particularly in the armies of Germany and Austria and their allies. The British, Americans and French tended to retain the rather longer-bladed bayonet: British and American bayonets were cut-and-thrust weapons; the French, with their famed Lebel bayonet, relied only on the thrust for effectiveness.

Bayonet design was reconsidered after World War I, resulting, for Britain, in a series of spike bayonets for thrusting only; in the French MAS '36 ramrod thrusting spike; and in the shorter knife bayonet employed by the Americans. Recent thinking on the design of the bayonet has brought it back closer to

Left: *Bougainville, Solomon Islands, March 1944; anticipating hand-to-hand fighting as he fires while advancing behind a Sherman tank against a Japanese-held position, a US infantryman has his long Garand bayonet fixed.*

Below: *Tanker of the German* Waffen-SS *(note collar patch), World War II, with a fighting knife tucked behind the lapel of his overalls; the grooved handle is typical of German knives.*

its origins of the 17th Century: a knife fixed to or stuck into the muzzle of a firearm. Thus, the modern bayonet tends to be a dual-purpose weapon, for use on the firearm or hand-held as a fighting knife. There have been many variations of late, with insulated wire-cutter bayonets, non-magnetic bayonets, bayonets with combination tool sets in the grip, and others. For the collector, bayonets offer a vast choice in the 20th Century alone.

FORMING A COLLECTION

Several points must be considered before the formation of a collection is undertaken. The obvious one is cost: how expensive are individual items likely to be, and how many such items will have to be acquired to build up a really worthwhile collection? Second is display and storage: a collector may form a large collection of swords only to find that lack of space does not allow them to be properly displayed; valuable items will require greater security, so the burden of insurance must be borne in mind. Third is the area and extent of specialisation: the type of weapon; the period; the country of origin. A collection of unrelated items of varying type and quality, although more easily and quickly assembled than a more specialised array, will give the collector little long-term satisfaction.

Purchases may be made from general antique shops and markets, militaria markets and fairs, specialist dealers, or at auctions. However, if the purchase of a particularly rare or expensive piece is contemplated, it is recommended that the collector should attempt to buy from an established and reputable dealer or through a leading auction house, who will guarantee the authenticity of the item.

Few examples of generally issued items may survive in near-mint condition—unless they come from unissued stock. Some collectors, indeed, prefer some evidence of use on the item, an interestingly realistic approach that applies also to some other aspects of militaria collecting, such as medals, where the requirement of mint condition is not as important as in coins. However, dress items are expected to be in fine condition: they should have had limited use and have been well cared for by their original owners.

A common bayonet may be purchased fairly cheaply; a scarce dagger or sword, especially if it once belonged to an eminent military figure, may be beyond the reach of some collectors. Items from a famous collection may fetch unrealistically high prices because of their provenance. Although price guides exist to assist the collector, these should always be consulted with understanding of the factors of rarity and condition.

CLEANING AND DISPLAY

Having purchased an edged weapon, one of the first things a collector may wish to do is to enhance its appearance by cleaning. This initial urge should be curbed until the necessary knowledge is acquired to carry out the work without risking damage to the weapon.

On polished steel surfaces, such as blades, abrasives should be avoided unless the blade is in a very poor state. Surface rust can be removed by using light oil and the edge of a copper coin (copper, being a soft metal, will not damage steel); only a small area should be worked on at a time. The same treatment can be used on "blued" surfaces. On a polished blade with etching, the state of the blade must be considered. If it is dull, but with only small spots of rust, white lime powder applied with a chamois cloth will restore a high polish. Bad patches of rust should be treated with either oil (as above) or with jewellers rouge applied with a cloth. Once the rust has been removed, the white lime powder may be used.

CARE OF GRIPS

Wooden grips, detachable on some weapons, were usually intended to be oil polished: a light application of linseed oil or good wax polish will usually be sufficient. If the grip is very dirty, then spirit should be used; it must be remembered that this method will remove the oil finish as well as dirt and must therefore be carried out over a period of weeks, allowing oil to oxidise before polishing and applying the next treatment with spirit.

Fishskin grips are best cleaned with dry powder, such as "Vim", and a toothbrush. Water and detergent will do the job; but the use of water may lead to the rotting of the wood grip beneath the fishskin and should thus

Above: *US paratrooper waits to jump in full combat equipment, c1945. Note that as well as a holstered pistol he is armed with a fighting knife—apparently a standard Case knife with a grip built up of compressed leather washers—carried in a scabbard strapped to his boot.*

Above right: *On the Western Front, c1916, a British Army sentry posted to give warning of gas attack (note the shell-case "alarm bell") has rested his rifle with 1907 Pattern bayonet fixed.*

be avoided. Naval swords have white Sun Ray fishskin and the use of cleaning powder will restore some of the whiteness. Other swords have grips covered in either leather or sharkskin which has been blackened. In these cases, careful use of "Shoemaker's Ink" or shoe dye will restore the covering. The use of dry powder with a tooth brush will also add some lustre to the wire binding. It is always better to underclean than overclean.

On guards and scabbards with nickel or chromium plating, automobile chrome polish will suffice. Brass guards and fittings which retain original gilt may be cleaned with a mild solution of ammonia to remove verdigris; this must be rinsed off and dried carefully afterwards. Brass hilts without gilding can be cleaned with a good brass polish.

Leather, particularly in scabbards, sheaths or frogs, may be cleaned with saddle soap, leather cream or even boot polish if the colour has faded. Dried leather should be brought carefully back to life by using a hide food; avoid rubbing. Leather scabbards on swords and bayonets may have rotted stitches. These can be re-sewn with waxed thread, using two needles and the original thread holes.

As an extension to cleaning, the collector must preserve his specimens. To protect blades from rust and figermarks they should be lightly wiped with a light gun oil (not the type used for lubricating purposes, but the type for maintaining gun barrels). With regard to storage, an even temperature is essential: avoid attics, where temperature changes can be considerable. Once cleaned and prepared, only occasional inspection and, if necessary, a further oiling, is required.

One of the most rewarding aspects of collecting is the display of the collection. The form that this will take is subject to cost and personal taste. Some collectors prefer to show edged weapons in shallow wall-mounted display cases, others prefer plan chests or chests of drawers. In either case, the display should embody protection that will help to maintain condition.

RECORDING AND RESEARCHING

As the collection grows, the serious collector will wish to document his items, recording details of purchase and the date, country of manufacture, type, blade length and any unusual features of the specimen. Such a list is also vital for insurance purposes; even more so when it is accompanied by photographs. Much of the pleasure of collecting lies in researching the origin and history of the item, by reference to some of the excellent and extensive books on the subject and by visits to museums. As one builds up a collection, one will often come into contact with other collectors, who are generally willing freely to give advice and share their knowledge. Visits to specialised auction sales will be rewarding: the items for sale are often catalogued in some detail and there will be an opportunity to handle them.

REPLICAS AND FORGERIES

With the increasing popularity of militaria collecting, particularly with regard to World War II items, a number of replicas or forgeries have come onto the market, sometimes offered innocently. Regulation swords will hardly ever be encountered, but Nazi daggers are produced in forms that may deceive the beginner. They are sometimes made up from an original dagger: parts are exchanged with a copy, thereby creating two part-original daggers! The study of blade inscriptions on Nazi daggers is necessary because of the possibility of spurious etched inscriptions of the kind encountered on genuine high-value presentation blades.

The rarer bayonets and fighting knives have also been copied or modified. For example, copies of scarce British No. 4 Mk I spike bayonets have been made from the No 4 Mk II bayonet by machining flutes in the blade and removing the Mk II stamp from the socket. The popular F-S fighting knife is reproduced in its Mk I form, but lacks the quality of the original Wilkinson Sword product.

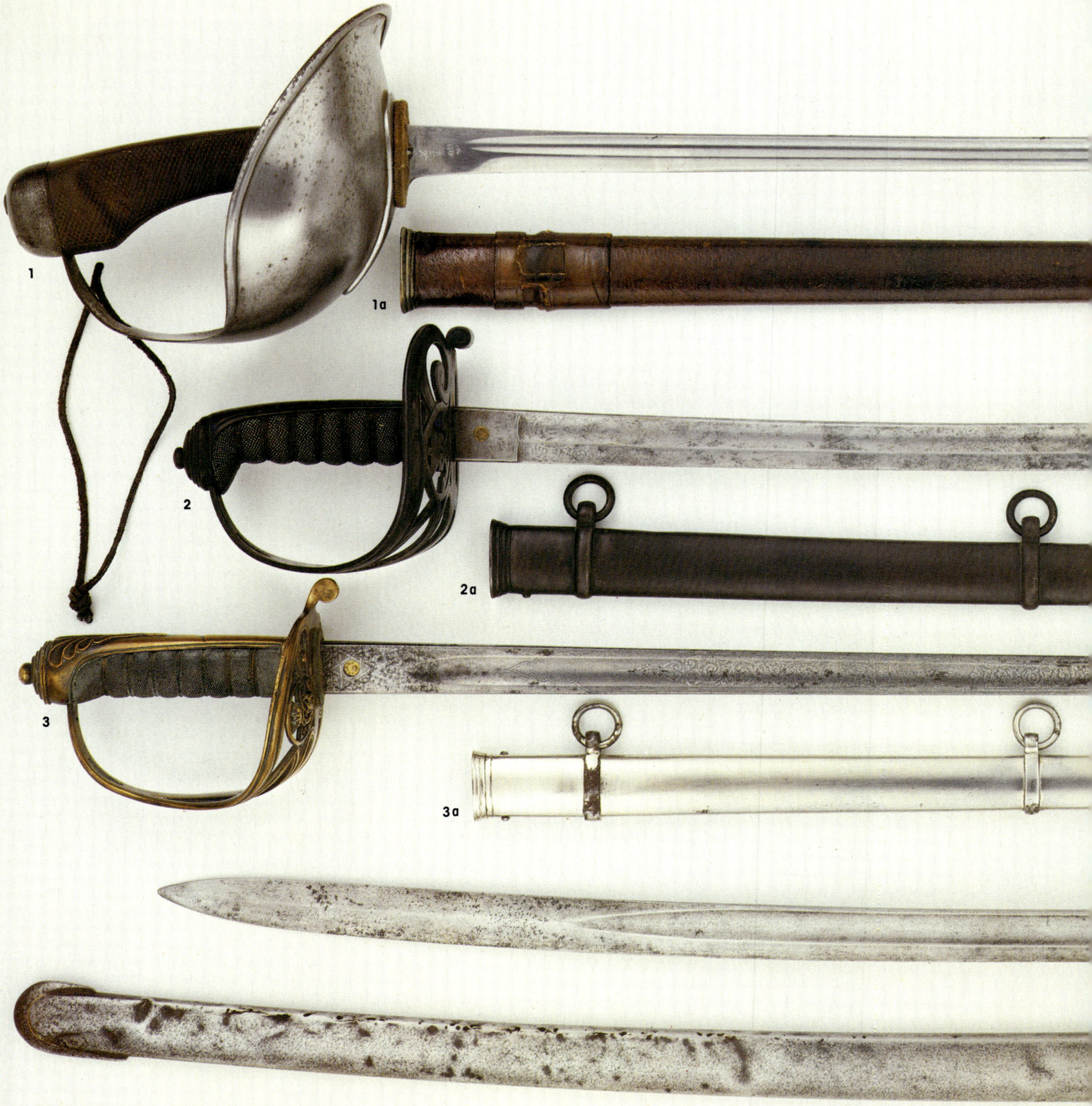

1 British Cavalry Trooper's Sword, 1908 Pattern. After exhaustive discussion and experiment by a War Office Committee convened in 1906, this sword was approved for the use of British cavalry troopers by King Edward VII on 2 July 1908. A change in the riding posture, the "seat", officially decreed for British cavalry in the early 1900s dictated that the new pattern should be designed primarily for use as a "thrusting" rather than a "cutting" weapon. Thus, purpose-designed, the 1908 Pattern is generally accepted to be one of the finest cavalry swords ever made, and it is ironic that it should have been adopted at a time when the sword as a practical weapon of warfare was obsolescent. Nevertheless, unlike most of the 20th-century swords shown in this book, which are basically ceremonial arms, the 1908 Pattern sword was used in action during World War I. The single-edged blade is 35in (889mm) long and is fullered to within about 9in (229mm) of the point. It has a steel, full bowl guard and a grip of hard, moulded, chequered rubber, with a thumb recess; the overall length being 42·75in (1086mm). The officer's version, the 1912 Pattern, is of similar dimensions, but has an etched blade, nickel-plated guard with engraved decoration, a decorative pommel and a wire-bound, fishskin-covered grip. Swords of this type are common.

1a This type of leather-covered wooden scabbard is usually found with the 1912 Pattern Cavalry Officer's Sword. The scabbard for the 1908 Pattern is similar to the one seen at (4a): all steel, either polished bright or painted khaki, with two suspension loops.

2 British Light Infantry Officer's Sword. The slightly curved, single-edged blade, 32·625in (829mm) long, is etched with foliage and

The broadswords traditionally carried by officers of Scottish regiments have detachable hilts: the basket-hilt shown, bearing the badge of the Gordon Highlanders, is worn with full dress and exchanged for a plain cross-guard with undress. The maker's number is visible on the blade of this example, made by Wilkinson, c1915.

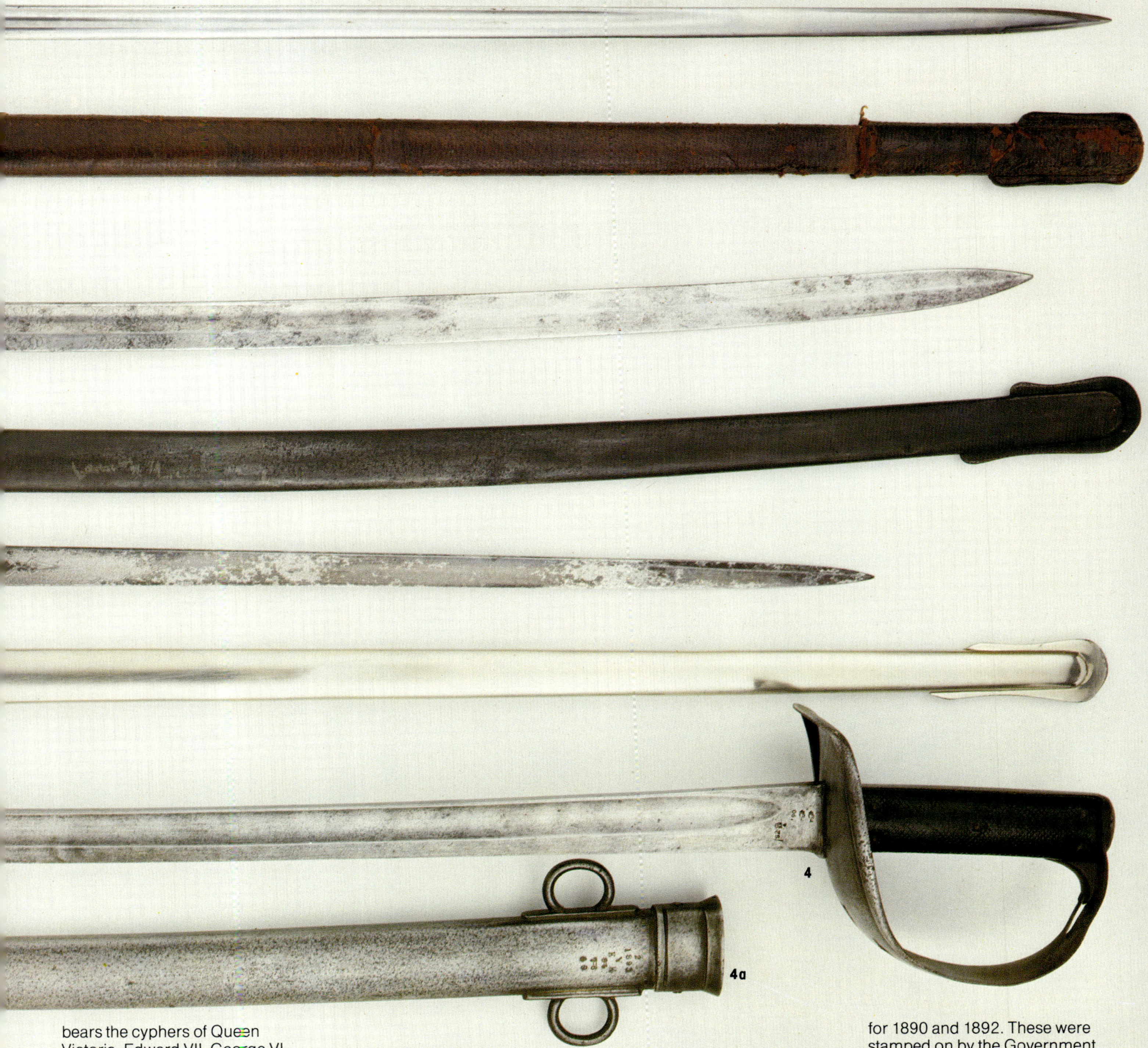

bears the cyphers of Queen Victoria, Edward VII, George VI and Elizabeth II; it is probable that examples with Edward VIII cypher exist – but such a feature would make this otherwise common sword very rare and valuable. The steel guard, a type known as a Gothic hilt, bears the Light Infantry's crown-and-buglehorn emblem on the hilt between the ferrule and knucklebow – the place usually reserved for the Royal cypher. The sword has a wire-bound, fishskin-covered grip and is 38·25 (972mm) long overall.

2a Scabbard for (2), steel, with two suspension loops and a dull finish. Such scabbards are more usually found polished bright, sometimes nickel-plated.

3 British Army Veterinary Corps Sword with George V cypher. It bears the name "Manton": this may be either the maker or the retailer – it is not always possible to distinguish which without reference to specialist texts. The gold disc within an etched Star of David at the ricasso (also called the forte) is a proof mark, indicating that the blade has passed bending tests to determine its quality and temper. The straight blade, 32·25in (819mm) long, is of the dumb-bell type (so called from the shape of a section taken through the fullers). The gilded brass triple-bar hilt is of the 1822 Pattern, with a folding guard on the reverse side to allow the sword to lie more closely to the body of the wearer. It has a wire-bound fishskin-covered grip and is 38·25in (972mm) overall. Such swords are fairly rare.

3a Scabbard for (3); nickel-plated steel with two suspension loops.

4 British Cavalry Trooper's Sword, 1885 Pattern. This type was still in use during the Boer War (1899-1902), but was to be replaced by the 1908 Pattern shown at (1). As may be seen, it is much more of a cutting weapon than its successor: the slightly-curved single-edged blade, 34·25in (870mm) long, has broad fullers. It bears the name "Mole": the manufactory associated with the name of Robert Mole was an important one from 1690 until 1921. Visible at the ricasso are acceptance marks for 1890 and 1892. These were stamped on by the Government armourer: normally, the original date of acceptance would be in full, sometimes with both month and year (eg, "10, 1902" for acceptance in October 1902), and further acceptance marks, on reissue, would be abbreviated (eg, "02" for 1902). The guard is sheet steel with rolled edges to increase its strength and has pierced Maltese cross decoration. The two-piece grip of chequered leather is secured by five rivets. Length overall is 39·625in (1006mm). Such swords will be fairly easily found by the collector.

4a Steel scabbard for (4). The two suspension loops are sited on either side of the scabbard near the throat, thus enabling it to be secured to the cavalryman's saddle.

1

3

1 British Heavy Cavalry Officer's Sword, 1887 Pattern. The very slightly curved blade, 36·25in (921mm) long, has broad fullers and is single-edged for most of its length but double-edged towards the tip. It is etched with decorative scrolls and has a proofmark within an etched Star of David at the ricasso. The guard is nickel-plated and its openwork scrolled decoration is of the type which has given hilts on which it appears the generic name of "honeysuckle" hilts, since the scrolls resemble the petals of that flower. It has a chequered backstrap, a rounded pommel and a wire-bound fishskin grip. The equivalent Light Cavalry (Hussars and Lancers, as opposed to Dragoons) sword has the same grip, a part-chequered backstrap and a triple-bar steel hilt; its blade is of similar design but with a length of 32·5in (825mm); thus the light cavalry sword is shorter and of generally lighter construction. The heavy cavalry sword has an overall length of 42·75in (1086mm). Note the leather sword knot: this feature, originally a functional item to prevent the user from losing his sword in the heat of action, is now purely decorative. The leather strap will often be found with a plaited "fir cone" terminal, as seen here. A leather sword knot is normally worn for field service purposes; a bullion-embroidered knot for dress occasions. On scabbards, a leather knot is worn with a leather scabbard; a bullion knot with a plated scabbard.

2 British Naval Officer's Sword with George V cypher. The single-edged blade, very slightly curved over its 31·5 (800mm) length, is etched with foliage scrolls. The triple-bar guard is of solid cast brass, gilded, and incorporates crown and anchor motifs of the service. The inner side of the guard folds to lock onto a stud at the throat of the scabbard. It has a brass lion-head pommel; the lion's mane flowing down to form the backstrap. The scales on the wire-bound, fishskin-covered grip are very pronounced. The sword, 37in (940mm) long overall, has a bullion dress knot: leather knots are not issued with naval swords. This is a fairly common weapon.

2a Scabbard for (2); black leather with gilded brass mounts and two suspension loops. In wear, the loops are attached to straps depending from the sword belt. Sometimes a hook depending from the belt allows the sword to be hooked up by a loop on the frog so that its tip clears the ground. Normally, the sword just touches the ground; hence the "drag" on the scabbard to prevent excess wear to the tip.

3 Scottish Officer's Basket-Hilted Broadsword with George V cypher. The straight, double-edged blade is 32·5 (825mm) long and has narrow fullers. It is etched with thistles and the badge of the Gordon Highlanders, with the gold-coloured hexagonal "best quality" proofmark of the famous maker Wilkinson at the ricasso. This indented brass hexagon as an identification of the company's

British Naval Officer's Dirk of regulation pattern; King's crown above anchor on langet and lion-head pommel; 23·25in (475mm) overall, 18in (475mm) blade.

highest-quality swords was adopted in 1905; the former proof-mark, round and also bearing the initials "HW", was retained for swords made for general sale or of "tailors' quality". (Note that the trade marks found on swords are not invariably those of the maker: some military outfitters who purchased large numbers of swords from a manufacturer were granted permission to have their own trade marks impressed on the weapons.) The plated basket hilt typical of Scottish swords is pierced with hearts in geometric patterns: such hilts are to be found with doe-skin liners. The pommel is detachable: a feature only found on Scottish swords, allowing the guard to be changed for the regimental pattern or "undress" pattern, a simple crosspiece. Basket-hilted swords like this one, which are rare, are worn only with full dress. The overall length of the weapon is 39in (991mm).

4 British Infantry Officer's Sword, 1892 Pattern; a very common sword of the pattern still carried. The straight blade, 32·5in (825mm) long, is of the dumb-bell type, with etched decoration that incorporates the George V cypher and, at the ricasso, the trademark of the Wilkinson Sword Company, the best-known of British makers. The royal cypher appears again above the scrolling of the sheet steel guard: the tails of the letters are just visible in the photograph. The wire-bound fishskin grip terminates in a rounded pommel and has a chequered backstrap; overall length is 38·5in (978mm).

4a Leather-covered wooden scabbard, with solid nickel throat, for (4). The leather frog is used to hang the sword from the two loops of the Sam Browne belt.

5 British Army Service Corps Officer's Sword with George V cypher. The single-edged, slightly curved blade is plain save for the Wilkinson maker's mark; it is broadly fullered and is 35in (889mm) long. The triple-bar hilt is of plated steel with rounded pommel and chequered backstrap (the latter feature being a distinguishing mark between this and the otherwise similar Royal Artillery Officer's Sword, which has a fluted top to the backstrap). This sword, easily found, is 41·5in (1054mm) long overall.

Swords of Imperial Germany and the Third Reich

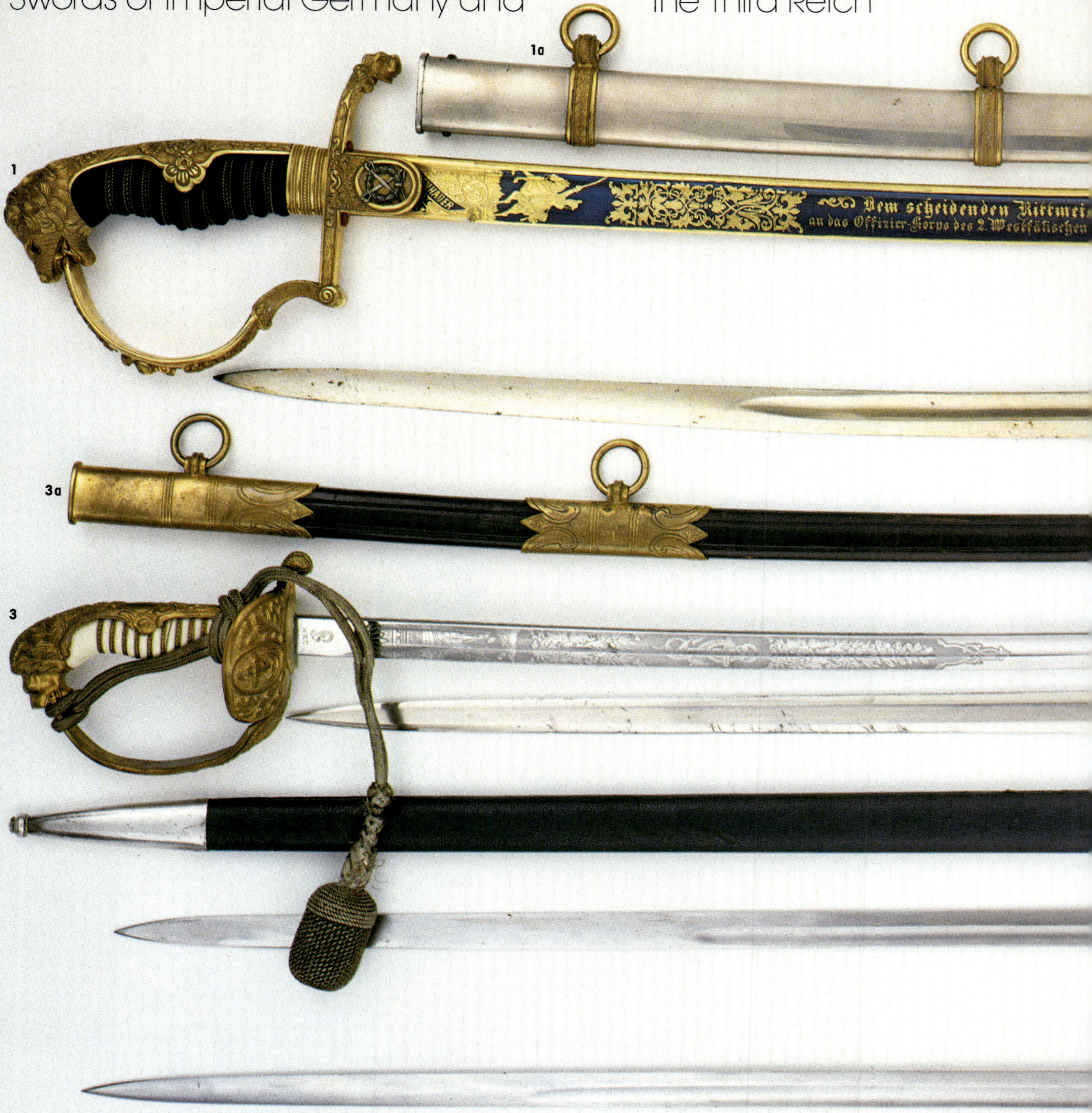

1 Imperial German Presentation Sword: a weapon of exceptionally high quality and a rare item. The damascus blade, 32·25in (819mm) long, is pipe-backed (ie, with a rod or "pipe" running along the back of the blade for added strength; sometimes called "quill-backed") with blued on gilt decoration that includes floral and scroll patterns, the Imperial insignia (towards the point) and a mounted cavalry officer (towards the hilt). The inscription tells us that it was presented to Rittmeister (Cavalry Captain) Berry as a remembrance of the officers of the 2nd Westphalian Hussar Regiment between 1880 and 1900: on the reverse of the blade are inscribed the names of the officers who contributed to the gift. The hilt, of gilt brass, is finely cast and hand-finished, with a lion-head pommel and quillon and a scrolled knucklebow. Overlaid on the langet are crossed swords and a wreath in silver. The grip is of silver-bound fishskin, and this magnificent sword is 38·5in (978mm) long overall.

1a Nickel-plated steel scabbard for (1), with gilt brass bands and suspension loops and an extremely large drag at the tip.

2 Imperial Germany Army Artillery Officer's Sabel (or Sabre; the German term is commonly used); a fairly rare item. The slightly curved single-edged blade, 31·75in (806mm) long, is etched with foliage framing a gun team. The knucklebow and backstrap are of steel and the wire-bound grip is covered with black celluloid. Length overall: 36·625in (930mm).

3 Third Reich Kriegsmarine (Navy) Officer's Sword, with the maker's mark of WKC (Weyersberg, Kirschbaum & Companie, Solingen) clearly visible at ricasso. The curved, fullered, single-edged blade is 33in (838mm) long and is etched with foliage, a fouled anchor and a warship: the only swastika markings are those shown on the warship's flags. The gilded brass hilt has a folding shellguard (ie, the guard is shaped like a shell; nearly all folding guards are in this shape) and the lion-head pommel has eyes

Hilt of a very rare German "Armistice" sword of World War I. The langet bears cast decoration of a "peace" dove and olive branch, with olives on the ferrule also.

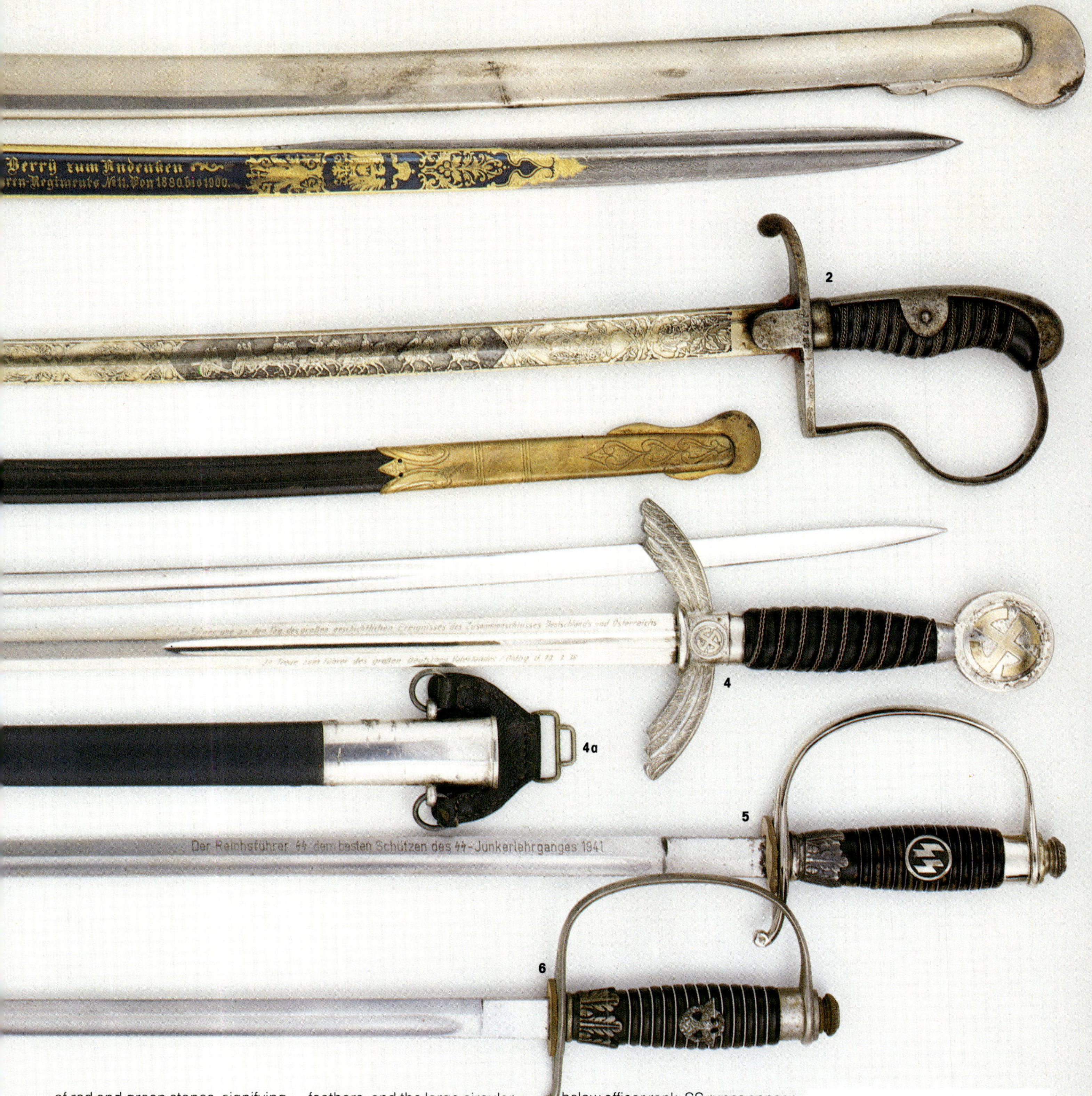

of red and green stones, signifying port and starboard. The wire-bound grip is covered with white celluloid. The sword, 38·125in (968mm) long overall, is shown with a silver-coloured dress knot (or portepee).

3a Black leather scabbard for (3); the brass mounts retain traces of the original gilding.

4 Third Reich Luftwaffe (Air Force) Presentation Sword. This pattern, fairly rare, like (3), was introduced in 1934. The straight, double-edged blade is 28in (711mm) long and bears an inscription commemorating the *Anschluss*, the forcible annexation of Austria on 13 March 1938, on either side of the short, narrow fullers. Gilt swastikas feature on the crosspiece, with downswept quillons symbolising feathers, and the large circular pommel. The wire-bound grip is covered in blue leather. Overall length is 35·125in (892mm).

4a Blue leather-covered scabbard for (4), with silver-plated mounts and blue leather suspender.

5 Third Reich Schutzstaffel (SS) Degen (Presentation Sword). The single-edged, fullered blade, 33·125in (841mm) long, has the etched inscription: *"Der Reichsführer SS dem Besten Schützen des SS-Junkerlehrganges 1941"* ("The Reichsführer SS Award for the best marksman in the SS Officers' Training Course, 1941"). It has an oakleaf ferrule, chromium-plated knucklebow and a blackened hardwood grip with inset SS runes. Overall length is 38·5in (978mm). On swords for those below officer rank, SS runes appear on the weapon's pommel rather than on the grip. A very rare item—and correspondingly valuable.

6 Third Reich Polizei (Police) Leader's Sword. Note the similarity to the SS sword (5). The blade, 29·5in (749mm) long, and the hilt are similar, but with a police badge replacing the SS runes on the grip. Overall length is 34·875in (886mm). This is a fairly rare item, though not so rare as (5).

Field-grey "attila" jacket of a German Lieutenant: the so-called "peace model" of 1915 which, it was planned, would be issued after Germany's victory in World War I. Some officers wore these uniforms unofficially while on home leave during World War I.

Swords of the Third Reich and Japan

1 Third Reich Army Officer's Sword, a fairly common item, bearing the maker's name of Paul Weyersberg, one of the various names associated with the famous and long-established German maker Weyersberg of Solingen, North Rhine-Westphalia, a town noted for its fine blades since the Middle Ages. The single-edged fullered blade, 31·25in (794mm) long, has etched floral decoration flanking (below) a helmet and crossed rifles motif and (above) the Wehrmacht eagle. The latter design is repeated on the langet, while the gilt brass hilt bears acorn and oakleaf relief decoration. The wire-bound grip is covered with black celluloid. Overall length: 36·375in (924mm).

2 Third Reich Bergbau (Mining Service) Sword, with the maker's identification "AC" (Alexander Coppel GmbH & Companie, Solingen). The plain, single-edged, fullered blade is 32.75in (832mm) long. The crossed mallet and hammer insignia of the Mining Service appears on the langet of the gilded brass hilt and the wire-bound grip is covered with black celluloid. (The celluloid-covered grip is a common feature of German edged weapons of the period.) Overall length of this fairly rare sword is 37·5in (952mm).

2a Brass-mounted black leather scabbard for (2).

Uniform tunic of an Officer of the General Staff of the Prussian Army, c1910. On the lower left breast is the embroidered cross of the Johannine Order of Prussia.

3 Third Reich Army Officer's Sword, of a similar pattern to (1), and also fairly common. The blade is 30·75in (781mm) long and is engraved with foliate panels incorporating the scrolled motto *"In Treue Fest"* ("Steadfast in Loyalty"). The eagle and swastika on the langet are smaller than on (1), and the hilt, with acorn decoration, is of white metal alloy which bears traces of a dull gilt finish.

3a Black-painted steel scabbard for the sword shown at (3).

4 Third Reich Army Sword of Lion-Head Pattern, with maker's mark of F.W. Höller & Companie, Solingen; a fairly rare example. The plain single-edged blade is 31·375in (797mm) long. The hilt is unusual: the ferrule and knuckle-bow are cast in one piece. The

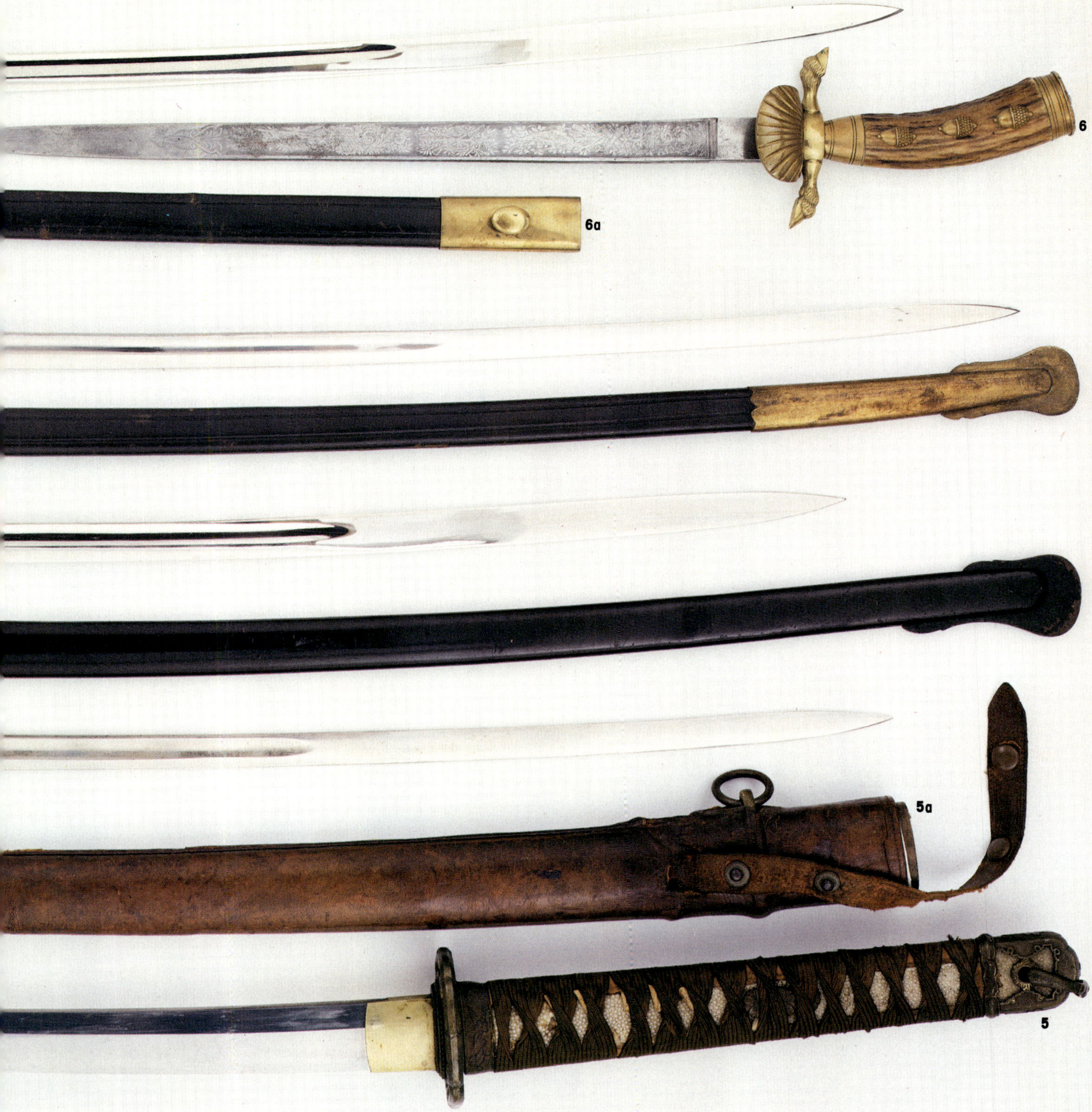

eagle with swastika on the langet is of the folded-wing type. Both quillon and pommel are of lion-head type, the latter with red stone eyes. The wire-bound grip is covered with black celluloid. Overall length is 36·5in (927mm).

5 Japanese Army Katana of World War II, a fairly common example with the maker's mark of a military arsenal. The single-edged blade is 29·375in (746mm) long: this classifies the sword as a *tachi* type—with a blade between 25-29·5in (635-750mm) long; the less common *wakizashi* had blades between 19·7-25in (500-635mm) long. Just visible in the photograph is the *yekiba*, the undulating line bordering the tempered edge. The cast-brass *tsuba*, disc guard, has imperial crysanthemum decoration—a motif repeated in the *menuki*, the ornamentation beneath the cord binding of the sharkskin grip. The double-handed hilt terminates in a *kabuto gane*, pommel, with a brass swivel loop. Overall length is 38·625in (981mm). Unlike most of the swords shown, the katana was meant for service: Japanese officers carried their swords into action and frequently used them. Japanese swords are perennially popular with collectors, and this is reflected in the prices of even the commoner arsenal products.

5a Wooden scabbard with brown leather cover for (5). Note the retaining strap to hold the sword in the scabbard: scabbards are sometimes found with spring clips for this purpose.

6 Third Reich Reichsbund Deutsche Jägerschaft (National Hunting Association) Sidearm. The single-edged blade, 406mm long, is etched with foliage and hunting trophies. The clam-shell guard, reversed quillons with hoof terminals, and pommel are all of brass. The grip, with three applied acorns, is stag-horn. This dress sidearm is a fairly rare item.

6a Brass-mounted leather sheath for (6), with oval frog stud.

Imperial German Navy Officer's Dagger. The wire-bound ivory grip has a crown pommel: officers who had served in World War I were allowed to wear these under the Third Reich. Note the rope motif of the upper suspension ring on the etched brass sheath.

Dress Daggers of the Third Reich

1 Third Reich Government Official's Dagger, for wear by members of the Nazi Foreign Office; a very rare item. The double-edged blade is 10·25in (260mm) long and the Nazi eagle forming the crosspiece is silver-plated, as is the stylised eagle's head forming the pommel. The grips are simulated mother-of-pearl. Overall length: 14·5in (368mm). The dagger worn by equivalent members of the Diplomatic Service is almost identical, but with the eagle's head on the crosspiece facing the opposite way.
1a Silver-plated sheath for (1), with pebbled panels and two suspension loops.
2 Third Reich Luftwaffe (Air Force) Dagger, First Pattern; introduced in 1934 for all officers and NCOs. The double-edged blade, 11·875in (302mm) long, bears the maker's mark of Paul Weyersberg, Solingen. The swastika at the centre of the drooping quillons is repeated on the circular pommel; both are silver-plated. The grip, with spiral wire-binding, is covered in blue leather. A fairly common specimen, with an overall length of 17·5in (444mm).
2a Blue leather-covered sheath for (2), with silver-plated mounts and hanging chains.
3 Third Reich Deutsche Luftsport-Verband (DLV; German Air Sports Association) Dagger. The 6·625in (168mm) blade bears the maker's mark of Josef Münch, Brotterode. The downswept, silver-plated quillons resemble those of (2), but the central black-enamelled swastika has straight bars. The grip is of wood covered in blue leather and the turned pommel is silver-plated. A fairly rare piece.
4 Third Reich Luftwaffe (Air Force) Officer's Dagger, 1937 Pattern. The double-edged blade, 10·125in (257mm) long, is etched with scrollwork and the Luftwaffe eagle, which is repeated on the aluminium crosspiece. The ball pommel, also of aluminium, bears a swastika and cast oakleaves. The ferrule is steel and the spiral-wire-bound grip is covered in orange celluloid. Overall length of this fairly common dagger is 15·25in (387mm).
4a Steel sheath for (4), pebbled, with oakleaf panel near tip; shown with hanging straps of silver-striped blue-grey fabric.
5 Third Reich Reichsforstamt (National Forestry Service), Senior Forester's Sidearm; a rare item of the cutlass type. The single-edged, 13in (330mm), blade is intricately etched with game beasts and dogs. The hilt, with a hoof-terminal to the quillon and a stag's head at the end of the knucklebow, is of gilded brass. The shell-guard has a pointer dog in relief. The grips are of ivory (junior ranks' sidearms had staghorn grips) with three acorns on each grip.
5a Black leather sheath for (5), with gilded brass mounts; note the acorn frog stud.
6 Third Reich Bahnschutzpolizei (Railway Protection Police) Dagger, 1938 Pattern. The double-edged, 10·25in (260mm) blade is similar to that of the

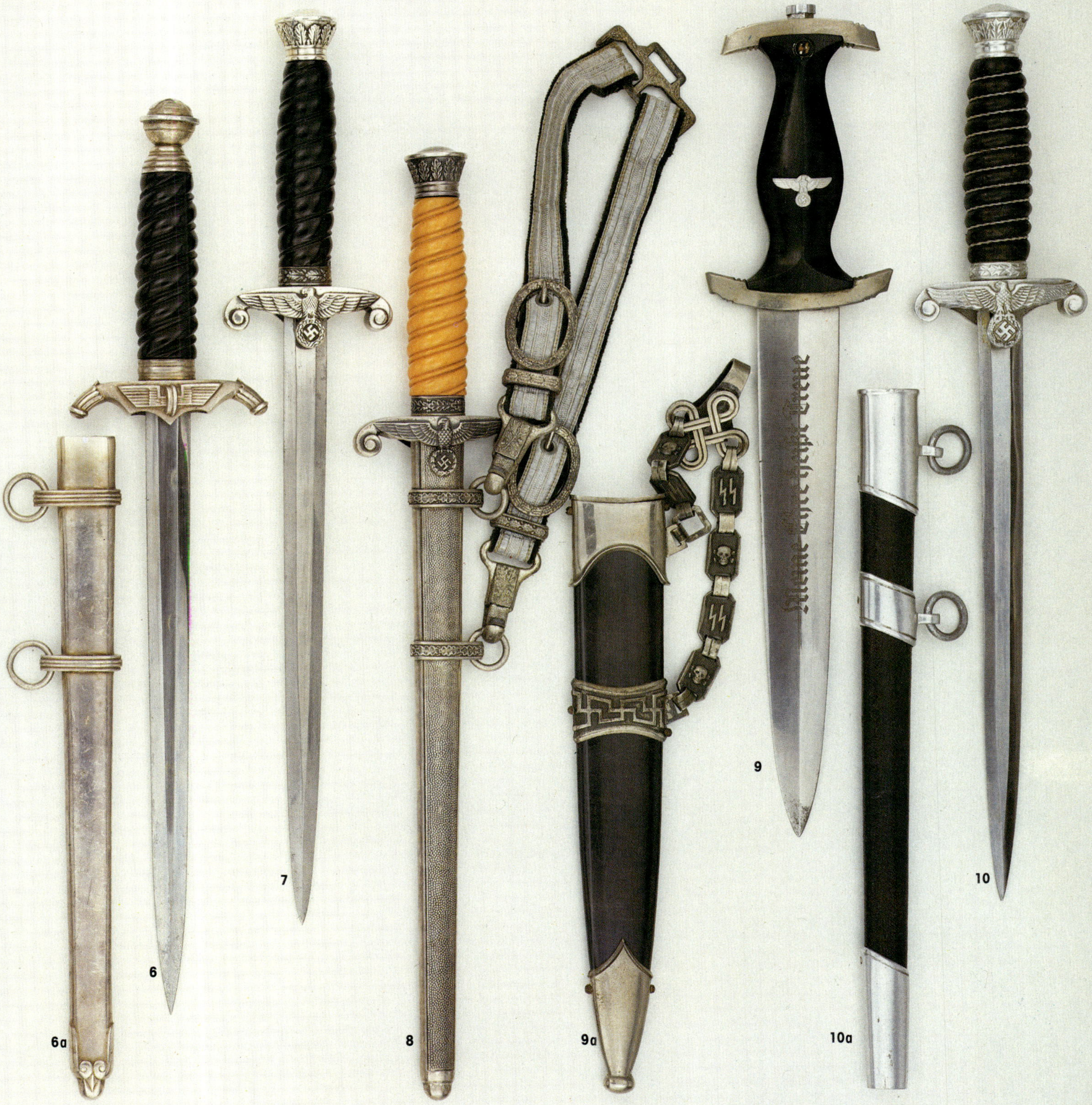

standard Army Dagger (8). The silver-plated crosspiece bears Railway insignia and there is a raised swastika on the dome of the silver-plated pommel. The grip is of black plastic, and the overall length of this very uncommon dagger is 15in (381mm).

6a Silver-plated sheath for (6), with double hanging loops.

7-8 Third Reich Heer (Army) Daggers. (8) is Standard Pattern; (7) is similar, but with the grip replaced by one of Bahnschutzpolizei type, see (6). Colours of standard grips vary from cream to dark orange. The double-edged blade is 10·25in (260mm) long. The crosspiece bears the Army eagle and the pommel is bordered with oakleaves. Overall length is 14·75in (375mm). The sheath shown at (8) has pebbled panels and two hanging rings and is equipped with hanging straps of silver-braided green fabric. Relatively common.

9 Third Reich Shutzstaffel (SS; Protection Department) Officer's Dagger, 1936 Pattern. The broad double-edged blade bears the motto *"Meine Ehre heisst Treue"* ("My honour is loyalty") and the black hardwood grip bears the eagle and swastika and, above, enamelled SS runes. Overall length is 13·625in (346mm). Such daggers, much prized by collectors, are rare and expensive.

9a Sheath for (9); the upper mount bears stylised swastikas and the hanging chain has links in which the death's-head emblem alternates with SS runes.

10 Third Reich Zolldienst (Customs Service) Official's Dagger. This rare arm has the same dimensions as the Army dagger (7-8) and a similar blade and pommel; the eagle on the crosspiece, however, has upturned wingtips. The grip is of wire-bound green leather, the colour identifying this as a Land Customs arm: the Waterways Customs dagger is exactly similar but has a blue grip and sheath.

10a Green leather-covered sheath for (10); with slanted alloy mounts and throat and two hanging loops.

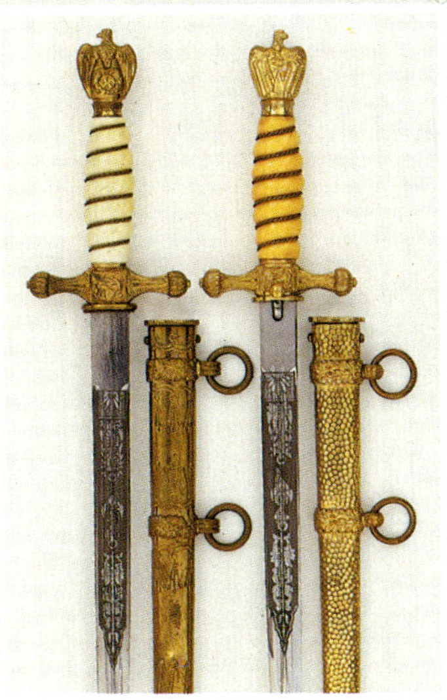

Third Reich Navy Officers' Daggers; common. (Left) Double-edged blade, 8·75in (222mm), with etched decoration. (Right) Reverse, to show securing catch at crosspiece, of variant with orange grip.

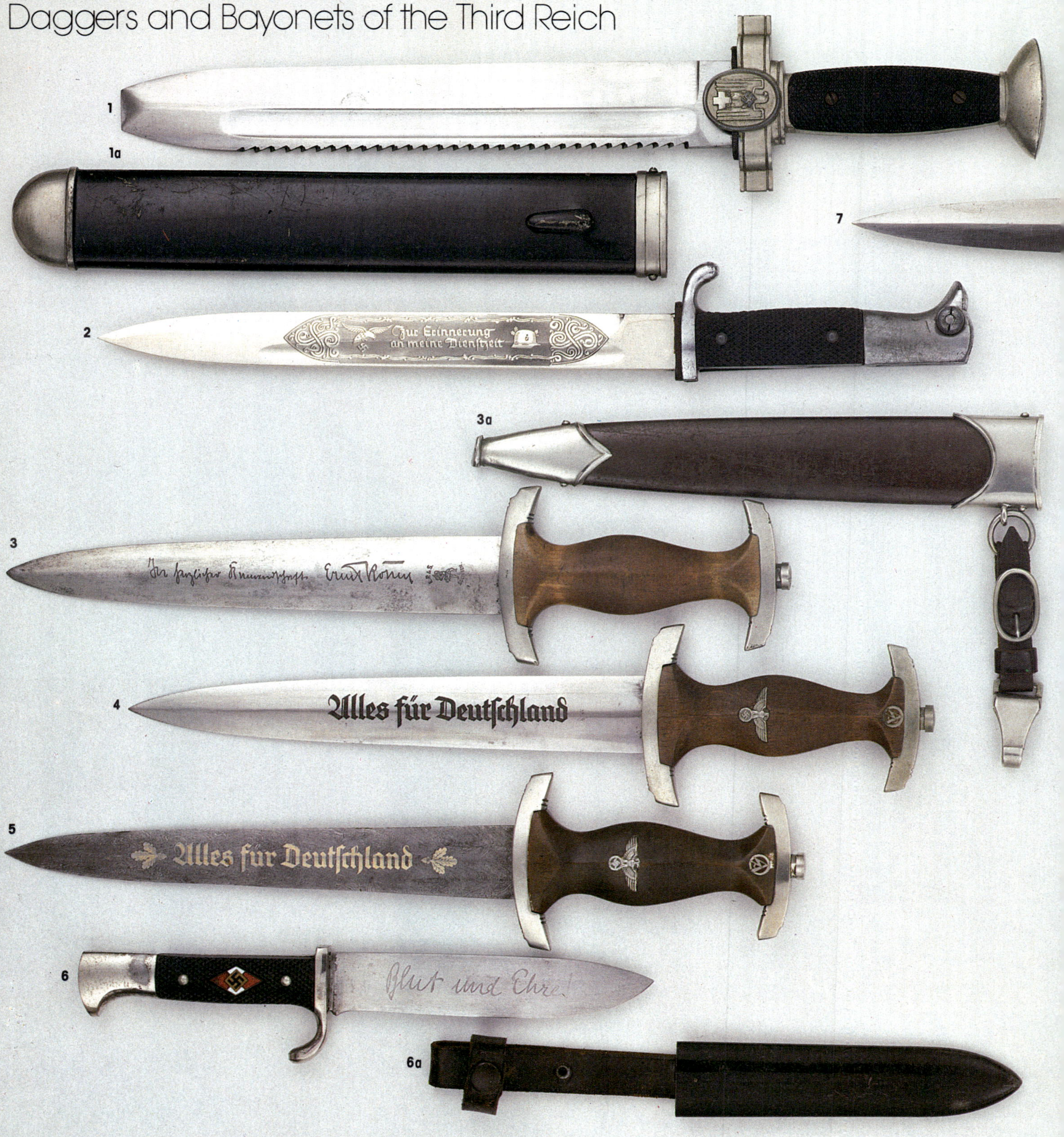

1 Third Reich Rotes Kreuz (Red Cross) Man's Dagger; a common type, adopted in 1938. The broad, broadly-fullered, single-edged blade, with saw back and chisel point, is 10.5in (267mm) long. The nickel-plated cast hilt has detachable grips: one chequered, as seen, and one plain. Although it appears to be a practical tool, this is a fairly fragile dress dagger. See also (7).

1a Black-painted steel sheath for (1), with plated mounts.

2 Third Reich Dress Bayonet. These arms, which are common, were made for private purchase by servicemen whose engagement had terminated. They were made with both plain and etched blades: the example shown bears the Luftwaffe eagle and steel helmet and the legend *"Zur Erinnerung an meine Dienstzeit"* ("In memory of my service time"). The single-edged blade is 9.875in (251mm) long, the crosspiece and pommel are plated, and the chequered black plastic grips are riveted to the hilt.

3 Third Reich Sturmabteilungen (SA; Stormtroopers, or Brownshirts) Dagger, 1933 Pattern. This is a rare piece, bearing on the reverse (as shown) the inscription *"In herzlicher Freundschaft Ernst Röhm"* ("In heartfelt friendship, Ernst Röhm"). More than 135,000 such daggers were awarded, but after the purge of the SA and the killing of Röhm, its leader, in June 1934, the vast majority had the inscriptions ground off by Hitler's order. The collector should beware of reproductions. The broad double-edged blade is 8.625in (219mm) long; the crosspiece and pommel are nickel-plated; and the grip is stained hardwood. Overall length is 14·625in (371mm).

3a Brown-painted steel sheath for (3), with nickel-plated mounts and hanging strap with catch.

4 Obverse of SA Dagger, see (3), showing the standard motto—*"Alles für Deutschland"* ("All for Germany") on the blade and the nickel eagle and swastika and enamelled SA emblem on the grip.

5 SA Dagger with damascus blade and motto in gold, flanked by oakleaf and acorn emblems. This is a very scarce item: the damascus blade is normally found only on presentation weapons for important leaders, although individuals could purchase such blades if they were prepared to pay a price about forty times greater than for the standard blade. The damascus process involves hammering together layers of steel and iron, polishing, and then dipping in acid, which attacks the two metals in different ways. It should be noted that there are artificial damascus blades, acid etched, which are worth little more than standard blades.

6 Third Reich Hitlerjugend (HJ; Hitler Youth) Knife; a common item. The single-edged, 5·5in (140mm) blade bears the motto *"Blut und Ehre"* ("Blood and Honour"); knives produced after 1938 have plain blades. The crosspiece and pommel are nickel-plated and the chequered black plastic grips, with enamelled HJ emblem, are riveted to the hilt. Overall length is 10in (254mm).

7a

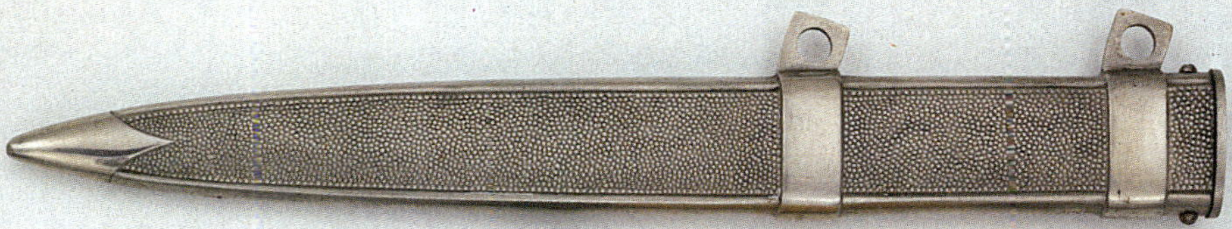

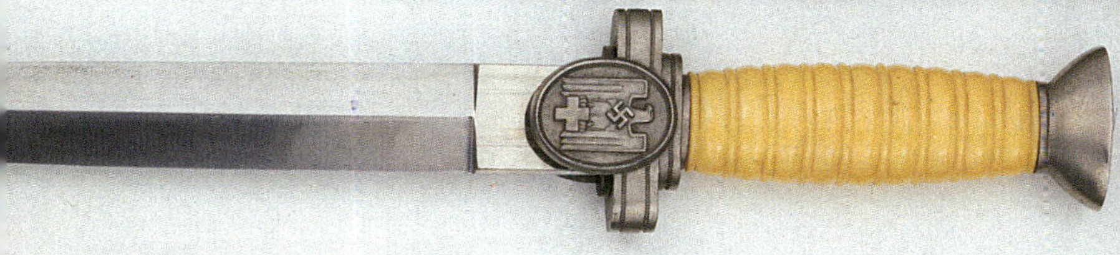

8a

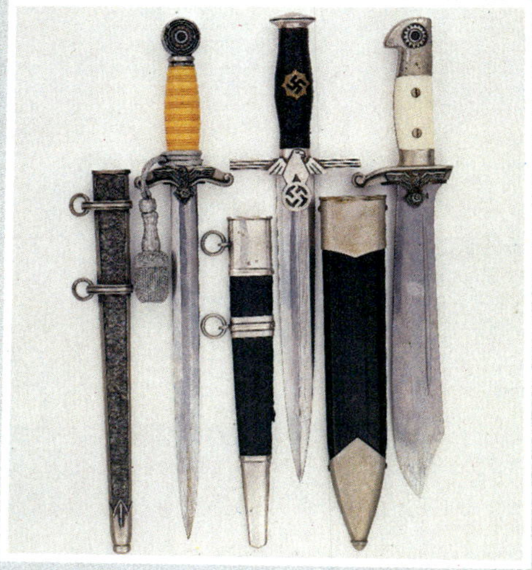

Dress side-arms of the Third Reich; for details see text in right-hand column, below.

8

9

9a

10a

10

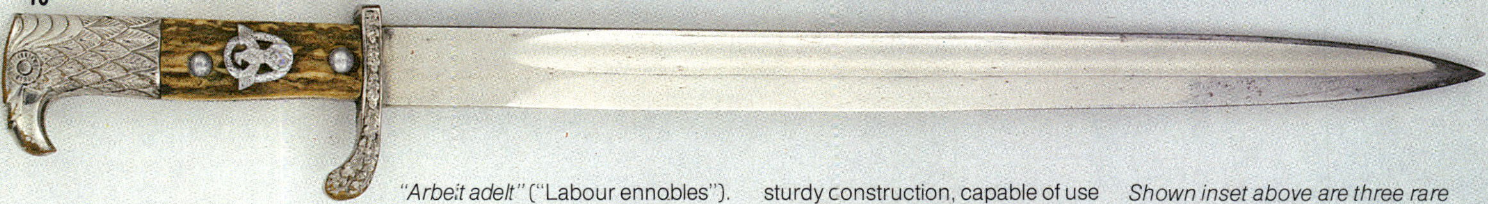

6a Black-painted steel sheath for (6), with leather belt-loop.

7 Third Reich Rotes Kreuz Officer's Dagger. This scarce item differs considerably from (1). It has a double-edged blade, 9·875in (251mm) long, and a nickel-plated crosspiece, with eagle and swastika langet, and pommel. The grip is yellow plastic (colours ranging from white to orange are found), with ten raised ribs.

7a Nickel-plated steel sheath for (7), with pebbled panels and two mounts for suspension straps.

8 Third Reich Reichsarbeitsdienst (RAD; National Labour Service) Officer's Dagger. The single-edged, narrowly-fullered, 10·25in (260mm) blade bears the motto *"Arbeit adelt"* ("Labour ennobles"). The crosspiece, with langet showing RAD badge, and eagle's-head pommel are silver-plated. The detachable grips are white plastic. Overall length: 15·875in (403mm). Six months' service in the para-military RAD, for construction work or agriculture, became compulsory in 1935 for all Germans between the ages of 17 and 25.

8a Silver-plated sheath for (8), with pebbled panels.

9 Third Reich RAD Man's Dagger. Reverse of heavy, narrowly-fullered, 9·87in (251mm) blade shown, with RAD triangle emblem, maker's mark of Carl Julius Krebs, Solingen, and number. The stag-horn grips are secured by two screws. Originally issued to all ranks of the RAD, this dagger—of sturdy construction, capable of use in the field—was issued only to ORs after the introduction of the Officer's Dagger (8) in 1937. Overall length of this fairly rare piece is 15·875in (403mm).

9a Black-painted steel sheath for (9); with nickel-plated mounts.

10 Third Reich Polizei (Police) Side-arm; a common item, basically a dress bayonet, but without a recess to fit the bayonet boss on a rifle. The fullers on the single-edged blade extend to the tip. The cross-piece bears oakleaf decoration and the eagle-head pommel has stylised feathers. Police insignia is applied to the riveted staghorn grips. Overall length is 19.5in (495mm).

10a Black leather sheath for (10), with plated mounts.

Shown inset above are three rare Third Reich daggers:
(Left) Technische Nothilfe (TeNo; Technical Emergency Help Corps) Leader's Dagger: 11in (279mm) long blade, with a nickel-plated crosspiece and pommel, silvered portepee and nickel-plated sheath;
(Centre) Reichsluftsschutzbund (RLB; Air Raid Protection League) Leader's Dagger: 9·625in (244mm) long blade, with silver-plated crosspiece and pommel, blue leather-covered grip and matching leather-covered sheath with silver-plated mounts.
(Right) TN Other Rank's Hewer; heavy, single-edged 9·75in (248mm) blade, nickel-plated crosspiece and pommel, white plastic grips and black-painted steel sheath with nickel-plated mounts.

1 British Bayonet, 1907 Pattern; for use on Short Magazine Lee Enfield (SMLE) Mark III rifle. The standard British bayonet of World War I, this remained in service until World War II and is very common. The single-edged blade is 17in (432mm) long; the detachable grips are walnut. Early examples have a crowned "ER" at the forte (ricasso) and a hooked quillon; most 1907 Pattern bayonets had quillons removed in 1913; later models have oil holes in pommel.
1a Steel-mounted black leather sheath for (1), with webbing frog; variations are found.
2 British Bayonet, 1903 Pattern; predecessor of (1), for use on SMLE Mark I, with 12in (305mm) double-edged blade. Since the SMLE rifle was shorter than the earlier Lee-Enfield, this bayonet soon gave way to the longer 1907 Pattern (1), and is fairly rare.
3 British Mark II Bayonet for No 5 Jungle Carbine, introduced 1944. The single-edged blade is 8in (203mm) long and there is a large muzzle ring to fit over the carbine's flash eliminator. The detachable beech grips are held by two screws; the Mark I Bayonet used only one screw. It could be fitted with a heavy wire knucklebow to become a fighting knife.
3a Black-painted steel sheath for (3), with brass throat and round frog stud. Same sheath used for No 7, No 9 and SLR bayonets.
4 US P17 Bayonet; single-edged blade, 17in (432mm) long. Note close similarity of this common arm to (1): the double grooves on the detachable walnut grips were necessary as a distinguishing mark, since the British and US bayonets were not interchangeable.
4a US steel-mounted green leather sheath for (4): the wire belt clip to fit US belts; the leather frog to fit British belts.
5 US Machete, World War II; Pattern No 1250, made by Collins of Hartford, Conn. The single-edged blade is 14·5in (368mm) long; overall length of this very common weapon is 19·5in (495mm).
6 US M1 Bayonet; introduced 1943 for use on M1 Garand rifle. The single-edged blade is 10in (254mm) long; overall length of this common weapon is 14·5in (368mm). The crosspiece, press catch and pommel are parkerised metal; detachable black composition grips are fitted.
7 German Model 1884/98 Bayonet; a common World War I example, with the maker's stamp of Alexander Coppel, Solingen. The single-edged, 10in (254mm) blade has angular fullers. Note 15 notches—possibly signifying victims—on walnut grip.
7a Green-painted steel sheath for (7), with leather frog. The white porte-pee with yellow stalk signifies the 1st or 9th Company, 3rd Battalion of an infantry regiment.
8 German Nahkampfmesser (Trench Dagger), World War I. A common example, with single-edged 5·5in (140mm) blade, straight crosspiece, and walnut grips which are marked with nine angled grooves and are secured by three rivets.

9 Bundeswehr (post-1945 West German Army) Flight Utility Knife, currently in service. This gravity knife, based on (13), has a retractable, single-edged, 3·375in (86mm) blade, a hinged spike, and chequered green plastic grips with a blackened alloy catch and housing incorporating a bottle opener. Length with blade extended: 8 75in (222mm).

10 German Ersatz (Emergency) Bayonet, World War I; used on Mauser rifles or, with adaptor, on captured arms. The single-edged blade is 12·25in (311) long; the hilt is cast brass, with a half muzzle ring secured to the tang by two rivets. Nearly 150 variations on the ersatz bayonet – inferior in quality to the standard arm, and now scarce – exist: they are found with brass or steel hilts and blade variations that include triangular socket types.

11 Third Reich Dress Bayonet with 7·75in (197mm) blade, as worn by NCOs; a common item. Nickel-plated crosspiece and pommel; black plastic grips. Shown with NCO's Troddel (side arm knot).

12 West German Flight Utility Knife, post-World War II; of lighter construction than (9), with a 4in (102mm) single-edged blade of stainless steel. A sprung flap secures the blade when retracted. Length extended: 9in (229mm).

13 Third Reich Flight Utility Knife, 1937 Pattern. These gravity knives (ie, designed to enable the user, with the aid of a spring catch, to free the blade with one hand only) were carried by Luftwaffe aircrew and by paratroops. See (14) for details.

14 Third Reich Flight Utility Knife shown with single-edged, 4·125in (105mm), stainless steel blade extended and pivoted spike partly open. The fittings are chrome-plated and the plain walnut grips secured by rivets. Such knives, 10·25in (257mm) long overall, are now quite rare.

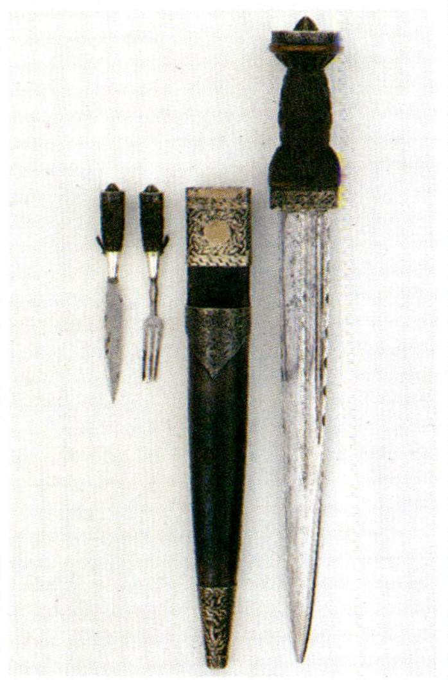

Scottish Officer's Dirk, c1915. The single-edged blade, 11·75in (298mm) long, has double fullers and a scalloped back. The silver-mounted grip is of hardwood carved to resemble basketwork. The leather-covered sheath, with silver-plated mounts, has a recess for knife and fork, which are fitted with securing clips. Overall length: 16·25in (413mm).

British and American Bayonets and Fighting Knives

1 British Machete, World War II; a common type, with a heavy, single-edged blade, 14·75in (375mm) long, dated 1943. The composition grips are secured by five copper rivets and are pierced with a hole for a retaining thong. Length overall is 20in (508mm). It was issued in a leather sheath (not shown here) with a belt strap secured by a metal stud.

2 British Jack-Knife, World War I type (although the example shown bears a Sheffield maker's mark and is dated 1938). The single-edged, 3·75in (95mm) blade has a deep nail groove; the double-edged can-opener has an opening stud; the utility spike is stamped with the War Department (WD) broad arrow. It has chequered black composition grips and a copper hanging loop. The overall length with the blade open is 8·375in (213mm). Like the jack-knives shown at (3) and (4), this knife is relatively common.

3 British Jack-Knife, post-World War II; dated 1949, with WD mark. It has five functions: single-edged, 2·75in (70mm) cutting blade; can/bottle opener; utility spike; screwdriver. Chequered black plastic grips and a steel hanging loop. Overall length: 6·25in (159mm).

4 British Jack-Knife, World War II; Sheffield-made and dated 1943. As (3), but with 2·5in (63mm) cutting blade and no bottle opener. Overall length, open: 6·125in (156mm).

5 British Fighting Knife (Push Dagger), World War I; made by Robbins of Dudley, a firm noted for the production of fighting knives for private purchase in 1914-18. This scarce item has a 4·875in (124mm) double-edged blade and a cast alloy "three-finger" grip, with a steel knucklebow. Overall length is 6·25in (159mm). It was supplied in a leather sheath.

6 British No 7 Bayonet, World War II; for use on No 4 Rifle and later marks of Sten Gun. It has an 8in (203mm) blade with a false edge (ie, a short cutting edge on the upper, non-cutting, edge, to aid penetration). A common item, but the design is unusual in that the socket can be swivelled through 180° to form a knife grip. The colours of the detachable composition grips vary from red to dark brown. Overall length: 12·375in (314mm).

7 British Fairbairn-Sykes (F-S) Fighting Knife, Second Pattern, World War II; made by Wilkinson Sword. The double-edged blade of flattened diamond section is 6·5in (165mm) long; the chequered brass grip and straight cross-guard are blackened. Overall length of the fairly rare item is 11·5in (292mm). Designed by Captains W.E. Fairbairn and E.A. Sykes, instructors in close combat, F-S knives were standard for British forces from 1941 onward and are carried by special forces to the present day.

8 British Military Issue Gurkha Kukri Knife, World War II. The single-edged blade is 13in (330mm) long and the knife, fairly rare, is 17·5in (444mm) overall. Unlike privately-made Gurkha knives, military issue

examples have walnut grips—not rosewood or ebony—and usually have official markings: the blade of this is stamped "M.I. 41 TT".

8a Leather-covered wood sheath for (8), with brass chape and detachable leather frog.

9 US LC/14B ("Woodman's Pal") Combination Fighting Knife and Machete, World War II; made by Victor Tool Company, Reading, Pa. The blued steel blade, 12in (305mm) long, has three cutting edges: for slashing, thrusting and hooking. The grip is built up of leather discs and the knucklebow is pressed steel. Length overall of this scarce item is 16·5in (419mm).

9a The sheath of the "Woodman's Pal", which is not shown here, was of green canvas, with zip fastening. It included space for storage of the two handbooks shown. One deals with the care and maintenance of the weapon, with instructions for sharpening. The other, shown open, gives user instruction, with emphasis on its value as a fighting knife. Note that the line illustrations show Japanese soldiers as the enemies to be cut down with this weapon.

10 US M5-I Bayonet; a pattern dating from the 1950s, for use with the M1 Garand Rifle. The single-edged, 6·625in (168mm) blade has a false edge and is parkerised (ie, with a dull sandblasted finish). The same blade was used in the M3 Fighting Knife and M4 Carbine Bayonet. The crosspiece has a stud rather than muzzle ring, to fit the late-model Garand. The detachable grips are of chequered black plastic. Overall length is 11.25in (286mm). A common item.

11 US Navy Air Corps Survival Knife; with maker's mark and dated 1945. The single-edged, folding blade is 10in (254mm) long and the overall length of this common item is 15·5in (394mm). The crosspiece is divided to allow for the folded blade, for which a simple lock catch is provided. Black plastic grips.

11a Sheet steel blade cover, to protect the cutting edge of blade of (11) when folded.

12 British No 4 Mark I Spike Bayonet; introduced in 1939 for use on the No 4 Mark I Rifle. This type, now scarce, marked the return of the socket bayonet last used by Britain on the Martini Enfield of 1895. The cruciform blade is 8in (203mm) long; overall length is 10in (254mm). The socket houses a spring-loaded catch that secures the bayonet to the rifle by means of two lugs.

13 British No 4 Mark III Bayonet; introduced 1943. A variation of (12), of simplified manufacture but with the same dimensions. Intermediate models were the Mark II, in which the original cruciform blade gave way to a circular section, and the Mark II*, which introduced two-piece construction. Both steel and plastic sheaths were issued.

14 British Mark II Sten Gun Socket Bayonet, World War II. The spike blade is 8in (203mm) long; overall length is 12in (305mm). The tubular socket is fitted with a flat spring locking clip. The sheath was that used for (13), but this bayonet, unlike (13), is now scarce.

German and Italian Bayonets and Fighting Knives

German Feldgrau *(field-grey) uniform tunic of World War I type. This is the Model 1907 tunic of an Infantry Lieutenant, with a bronze wound badge.*

1 German Mauser Bayonet, 1898 Pattern. This fairly scarce bayonet has a fullered 20·5in (521mm), pipe-back (or quill-back) blade with a spear point. The quillon sweeps up towards the pommel, which has an S-curve to butt against walnut grips with nine angled grooves. The weapon is sometimes found with a one-piece walnut grip. Overall length is 25·375in (645mm).

1a Steel-mounted leather sheath for (1), with a black leather frog bearing the portepee of the 4th or 8th Company of the 2nd Battalion of an infantry regiment. The decorative tassel was known as a portepee when worn by an officer; a *Faustriemen* when worn by a cavalryman or rifleman below commissioned rank; and a *Troddel* when worn by a non-commissioned member of any other arm of service.

2 Third Reich Feuerwehren (Fire Department) Junior Subordinate Ranks' Dress Sidearm. The nickel-plated, fullered, single-edged blade is 9.75in (248mm) long—the length sanctioned for Junior Subordinate Ranks of the Fire Department—and bears the maker's mark of WKC (Weyersberg, Kirschbaum & Companie, Solingen). The cross-piece, reversed quillons and pommel are all plated, and the chequered black plastic grips are secured by two rivets. Overall length: 14·5in (368mm).

3 Third Reich Feuerwehren Senior Subordinate Ranks' Dress Side-arm; like (2), a common item. The plated, single-edged blade, 7·875in (200mm) long, has narrow fullers and a pronounced false edge; the hilt is similar to (2). Overall length: 12·75in (324mm).

3a Black-painted steel sheath for (3), with black leather frog and Feuerwehren portepee.

4 German Pioneer Bayonet, Model 98/05, called the "Butcher's Knife Bayonet". This saw-backed arm, now quite rare, caused an outcry over "Hun beastliness" in the Allied Press during World War I (the British Army had issued saw-backed bayonets during the 19th Century, but had abandoned the type by 1903). The 14·5in (368mm) blade, swelling towards the tip, has a double row of saw teeth. The hilt resembles that of (1), with a longer quillon, but is usually found with a flashguard along the back of the grips to protect the wood from

muzzle blast. Overall length is 19·75in (502mm).

5 German Mauser Bayonet, Model 98/05; a standard World War I type without saw-back. Dimensions of this common specimen as (4); the flashguard on the back of the grips is just visible.

5a Steel-mounted leather sheath for (4) and (5); all-steel versions are more common.

6 German Mauser Bayonet, Model 84/98; the standard German Army bayonet of World Wars I and II. See also (11). The blued, single-edged, fullered blade is 10in (254mm) long. The steel crosspiece and pommel are blued and the detachable walnut grips are fitted with a flash-guard. Overall length is 15·25in (387mm). This pattern is common; a saw-back version—dating from World War I is less easily found.

7 Italian Carcano Bayonet, Model 1891. Single-edged over most of its length, the fullered blade becomes double-edged towards the tip. The crosspiece, straight quillon with ball terminal, and pommel are of blued steel. The hardwood grips are secured by two rivets. Length overall is 16·25in (413mm). A common item.

7a Blued steel sheath, with ribbed panels, for (7), with screw securing frog stud. Brass-mounted leather sheaths may be encountered.

8 German Nahkampfmesser (Trench Dagger), World War I; a privately purchased weapon of a common type, based on a hunting knife. The double-edged blade is 6.25in (159mm) long. The crosspiece, with reversed quillons, is steel; the staghorn grip has a nickel ferrule.

8a Nickel-mounted leather sheath for (8), with integral belt loop.

9 German Short Saw-Back Bayonet, Model 1898; dimensions as (6). The single-edged, fullered blade has a double row of saw-teeth. The swept-back quillon resembles (4) and (5), but the pommel is of a more pronounced "bird's-head" type. The chequered leather grips are secured by three rivets.

9a Black-painted steel sheath for (9), with black leather frog.

10 Italian Fascist Milizia Voluntaria Sicurezza Nationale (MVSN; Voluntary Militia for National Security) Dagger, Second Pattern; intended both for dress wear and for use as a fighting knife. Many were carried in World War II, but they are now quite rare. The 8in (203mm) blade is part double-edged. The steel crosspiece has a back-swept quillon and the hardwood grip is secured by two rivets and pommel capping. Overall length is 12·375in (314mm).

10a Blackened steel sheath for (10), with embossed fasces symbol of the Italian Fascist Party and "M.V.S.N.". There is a brass insert at the throat.

11 German Mauser Bayonet, Model 84/98, Third Reich type. Otherwise similar to (6), this bayonet has a blued blade and brown plastic grips with seven angled grooves. This type of grip was typical of the bayonets produced during the Third Reich period; examples with wooden grips may be found.

European and Canadian Bayonets and Fighting Knives

1 Finnish Bayonet Model 1926; also called the 1927 Pattern. This fairly common bayonet has a single-edged, fullered blade, 11·875in (302mm) long, with a profile similar to that of the Italian Carcano Bayonet Model 1891 (see 7, *page 37*). The tang (that part of the blade running into the hilt) swells below the steel crosspiece, which has a forward-swept quillon. The maker's name, Hackman, is at the ricasso (reverse). The steel pommel makes an angled joint to the walnut grips, which are secured by two steel rivets. Overall length of the arm: 16·25in (413mm).

1a Blued-steel sheath for (1), with knobbed chape and frog stud.

2 Canadian Ross 1910 Pattern Bayonet; generally known as the Mark II Bayonet and used with the Ross Mark II Rifle of World War I. The single-edged blade is 10·125in (257mm) long and has a long taper to the point: the Ross Mark I Bayonet, with a more rounded point, had been found to require excessive pressure for penetration. The crosspiece is of heavy steel and the angular steel pommel is stamped "ROSS RIFLE CO/ QUEBEC./PATENTED. 1907". The walnut grips are detachable. Overall length: 14·75in (375mm). The Ross Rifle, with its "straight pull" action, was found unsuitable for trench warfare and withdrawn in 1916; some were issued to the British Home Guard early in World War II, and it is fairly common.

2a Brown leather sheath for (2), incorporating an integral frog and a metal-studded chape.

3 German Fighting Knife (*Nahkampfmesser*) of the inter-war period, with a 6in (152mm) blade, double-edged towards the point; made by Anton Wingen Jr, Solingen. A non-functional press-stud forms the "eye" of the eagle's-head pommel; the grips are chequered black plastic. A fairly common knife, with an overall length of 10·35in (264mm).

3a Black-painted steel sheath for (3), with leather belt-loop and securing strap with press-stud.

4 British Fairbairn-Sykes Fighting Knife ("Commando Knife"), Third Pattern, of post-World War II manufacture. The double-edged blade is 7in (178mm) long and is polished, establishing this as a privately-purchased example: issue knives have blackened blades. The crosspiece is brass and the grip, with the concentric rings characteristic of the Third Pattern, is cast alloy. A brass pommel nut secures the tang, which extends through the grip. Overall length is 11·5in (292mm). Fairbairn-Sykes knives of various types are quite common: the original British manufacturer, Wilkinson Sword, produced more than 56,000 Second Pattern knives alone to British and Allied government orders. A near-identical knife was made in the USA by the Case Cutlery Company for OSS personnel.

4a Leather sheath for (4), with brass chape and brass-mounted leather tabs to allow the sheath to be sewn on to a tunic.

5 Spanish Mauser Bolo Bayonet; a

fairly common bayonet used on Spain's Mauser service rifles of the earlier 20th Century, with the cutting edge of the heavy, fullered, single-edged blade ground to the machete-like "bolo" shape. Maker's mark, "FN TOLEDO", at ricasso. The crosspiece, with a flattened ball terminal to the quillon, and the bird's-head pommel are blued, and the latter is fitted with a detachable adaptor that allows the bayonet to be used with German Mauser rifles. Overall length, with adaptor: 15in (381mm).

5a Blued-steel sheath for (5), with knobbed chape and oval frog stud.

6 Swedish Mauser Bayonet, 1898 Pattern; used on Swedish Mauser service rifles of the earlier 20th Century. Again a fairly common weapon, this is an all-steel bayonet. Its double-edged blade, 8·187in (208mm) long, has narrow central fullers. It has a cutaway crosspiece and the chequered grip, of round section, has an angled fixing catch. Overall: 13in (330mm).

6a Centrally-fluted steel sheath for (6), with an exceptionally large leather frog incorporating a securing strap that fastens over the frog stud.

7 German Trench Dagger of World War I; a fairly common type, double-edged towards the point of 5·75in (146mm) long blade. An acceptance stamp is visible just below the steel oval crosspiece. The slab-sided walnut grips, with nine angled grooves (the standard number on weapons of this kind), are secured by two rivets. Overall length: 10·875in (276mm).

7a Blued-steel sheath for (7), with leather belt loop and securing strap.

8 German Trench Dagger of World War I; one of the most common types. The single-edged blade, 5·875in (149mm) long, bears at the ricasso the maker's mark "ERN/WALD, RHEINL", showing it to be the product of C. Friedrich Ern, Ern-Rasiermesserfabrik, Solingen-Wald, a noted manufacturer of trench knives during World War I. It has a steel oval crosspiece and the walnut grips, with nine angled grooves, are secured by three rivets. Overall length: 10·75in (273mm).

9 German Fighting Knife of World War; a fairly common example of a good-quality knife. A blade length of around 6in (152mm) is usual in these weapons: the knife shown has had its single-edged blade re-pointed, thus shortening it to 5·25in (133mm). The back-swept quillon and bird's-head pommel, with a non-functional press-stud forming the "eye", are common in weapons of this kind; the stag-horn grips, found on better-quality knives, are secured by two rivets. Overall length: 10in (254mm).

10 German Trench Dagger of World War I; a further variation on the type shown at (7) and (8): many such variants are to be found. The steel oval crosspiece is smaller than those of the earlier examples; the walnut grips are more squarely cut and the nine angled grooves run in the opposite direction. Blade length: 6in (152mm); overall length: 10·75in (273mm).

Bayonets: France, Japan, Brazil, Czechoslovakia

1 Czechoslovakian Mauser Bayonet Model 33/40 of World War II; for use on the Czech Model 33 Carbine (which was taken into German service during World War II as the Gew 33/40 Rifle). This is a fairly common bayonet, but somewhat unusual in that the blade, which is 11·75in (298mm) long, has its cutting edge uppermost. The crosspiece, muzzle-ring and bird's-head pommel are, like the blade, of blued steel, and the walnut grips are detachable. Overall length: 17in (432mm).

1a Blued-steel sheath for (1), with knobbed chape and simple metal frog loop.

2 Swiss Schmidt-Rubin Model 1931 Bayonet; a well-finished, high-quality weapon, and fairly common. The centrally-ribbed, double-edged blade, with a profile closely resembling those of the British Pattern 1888 and Pattern 1903 Bayonets (see 2, *page 32*), is 12in (305mm) long. At the ricasso is the manufacturer's mark "WAFFENFABRIK/NEUHAUSEN": Neuhausen Rhinefalls is the home of the famous SIG Swiss Industrial Company, prominent in the manufacture of arms since the later 19th Century. The crosspiece (which bears a serial number, marking this as an issue item) and bird's-head pommel are of polished steel and the walnut grips are secured by two rivets. Overall length: 17in (432mm).

2a Blued-steel sheath for (2); with leather strap which is attached to the frog loop.

3 Japanese 30th Year Type (1897 Pattern) Bayonet (*Juken:* "rifle sword"); the standard Japanese bayonet of two World Wars, designed for use on the 30th Year Type "Arisaka" Rifle of 1897. Fairly common, this is a heavy and sturdy weapon: bayonet fighting received considerable emphasis in Japanese Army training and, as reference to contemporary photographs will show, Japanese infantrymen almost always went into action with bayonets fixed. The fullered, single-edged blade is 15·75in (400mm) long; it bears at the ricasso the triple-circle mark of the Nagoya Arsenal, together with the mark of the Nagoya sub-plant where it was produced. The hooked quillon on the blued-steel crosspiece is intended to trap and break the blade of an opponent in close-quarter fighting. The walnut grips are detachable. These bayonets are fairly common: 30th Year Type Bayonets lacking the hooked quillon and of inferior material, manufactured late in World War II, are also to be found. Overall length: 20·125in (511mm).

3a Blued-steel sheath for (3), with frog loop.

4 French Lebel Bayonet Model 1886/93/16. This weapon, introduced in 1916, is very similar to the bayonet designed for use with the Model 1886 Lebel Rifle, but lacks the hooked quillon of the original pattern—see (8). It is unusual in several ways. The 20·5in (521mm) long blade is formed by cutting four deep grooves into a circular block, thus having a cruciform section—as have (7) and

(8). The muzzle ring incorporates a locking device. The all-metal grip, found in both nickel and brass, is detachable on some examples. Overall length: 25·125in (638mm).

4a Blued-steel tapered sheath for (4), with knobbed chape and steel frog loop (see 7a).

5 Brazilian Mauser Bayonet Model 1908; a fairly common weapon, used with the Brazilian 7mm Mauser Rifle, Model 1908, which differed basically only in calibre from the German Gew 98. The bayonet itself is of German manufacture: it bears the maker's mark "W.K & CIE/SOLINGEN" (Weyersberg, Kirschbaum & Company) at the ricasso. The fullered, single-edged blade is 11·75in (298mm) long. The crosspiece, with a hooked quillon, and pommel are of polished steel; the walnut grips are secured by two rivets. Overall length of the weapon is 17in (432mm).

5a Black leather sheath for (5), with brass chape and locket (the reinforced section incorporating the throat). Note the serial number on the frog stud, indicating that this was an issue rather than a purchased item.

6 French Mousquetoon Bayonet Model 1892; designed for use with the 8mm Model 1892 Mannlicher Berthier Mousquetoon (Carbine). This is again a fairly common weapon, but the single-edged blade—15·75in (400mm) long and of cruciform section like (4), (7) and (8)—has two unusual features: the back is grooved to a length of 6in (152mm) from the tip, and at the hilt end of the fullers are two recesses to accommodate the two internal securing springs in the sheath. Another unusual feature is found on the pommel, which has a central hole to accommodate a fixing bar on the carbine. The hooked quillon marks this as an early example: it was officially removed in 1915. The hardwood grips are secured by two rivets. Overall length: 20·5in (521mm).

7 French Lebel Bayonet Model 1886/35; the final version of the Model 1886—see (8)—although this example retains the hooked quillon. The major difference is that in 1935 most Lebel bayonets had their blades shortened, as seen here, to 13·5in (343mm). The grip is nickel. Overall length of the weapon is 18·125in (460mm).

7a Converted sheath for (7), with leather frog—a more complex type than most, used with all Lebel bayonets—showing the way in which the securing strap passes through the frog loop.

8 French Lebel Bayonet Model 1886; the original type—see also (4) and (7). The round press catch on the pommel is chequered. The cruciform blade is 20·5in (521mm) long. The weapon has a hooked quillon and the grip is nickel. Overall length: 25·5in (648mm). Note that for obvious reasons—bayonets must be sharpened!—some variation in length will often be found between bayonets of identical pattern; this is broadly true of all edged weapons.

8a Blued-steel sheath for (8), with knobbed chape and frog loop.

Metal Insignia

Bob Cayley, Tony Walker, Robert Wilkinson-Latham

Left: *Field Marshal Montgomery, seen with US Generals Patton and Bradley in 1944, adopted the "trademark" of two badges on his beret: that of his rank and that of the Royal Tank Regiment.*

Metal insignia, which basically break down into badges, buttons and shoulder titles, are among the types of militaria which are most easily found in quantity and, generally speaking, quality. There are, of course, rarities in the field, but since all nations' fighting forces have their own badges and buttons, with many of the designs remaining virtually unchanged for many years, it is obvious that in most cases there will be a considerable number of examples of any one badge design still in existence.

The practice of using badges to distinguish one unit or arm of service from another was in use by the 17th Century or even earlier, but with the general advent of standing armies it became more important for servicemen to be able to distinguish one unit from another and, of course, from the enemy. As other sections of this book demonstrate, such distinguishing marks took many forms, notably head-dress and uniform colours (as well as flags and banners, which fall outside our survey); but, certainly in British service, perhaps the most basic and significant marks of unit individuality are the badge and button. In many armies, "general service" buttons for Other Ranks were issued in the late 19th and early 20th Centuries: although soldiers might accept these, a "general service" cap badge would be unthinkable for men of most units.

Not all countries have the same rich tradition of unit badges as Great Britain, where the badge is often a design that may refer back two or three hundred years to the unit's origin, incorporating also its name and major battle honours. It is inevitable, therefore, that our selection from the vast field of metal insignia should include much British material (*pages 46-55*), for its attraction is such that it is sought by collectors all over the world. However, American badges (*pages 60-61, 64-65*) and French badges (*pages 62-63, 64-65*), though not so numerous as British, make up in quality what they lack in quantity, while German badges (*pages 66-67*), particularly the combat awards of the Third Reich, are increasing in popularity—and in expense—among modern collectors.

MODERN BADGES & BUTTONS

Our representative selection of badges and buttons spans the period from 1900 to the present day. Since most armies had formulated their designs by around the beginning of this period, the novice may think that the field is a fairly restricted one. This is most definitely not the case.

On mobilisation in 1914, the services involved put into storage the full dress they had hitherto worn—and it was not generally issued again. Between the two World Wars, Officers wore full dress on certain occasions and some military bands managed to obtain full dress items for wear; but we are mainly concerned here with the cap badges worn on field service caps, side caps, forage caps and berets; the buttons worn on service dress; and the badges worn by the combat soldier, sailor or airman. It may be noted that full dress badges and helmet plates although larger, often have the same central design and many of the other features of the smaller cap badge.

Badges and buttons have been made in various metals in this century. In British service, the most common are brass, for Other Ranks, and gilt for Officer's full dress caps, side caps and berets or bronzed for Officers' service dress. Other badges, the result of regi-

Above: *Men of the Prince of Wales's Leinster Regiment (Royal Canadians) in c1905. The regiment's cap badge bore the Prince of Wales's plumes; it was one of the Southern Irish regiments disbanded in 1922 following the establishment of the Irish Free State.*

Left: *A German prisoner-of-war wears the Infantry Assault Badge on his left breast pocket; Normandy, June 1944.*

Right: *NCO of the Regiment de Marche du Tchad, serving with the Free French in the UK, 1944, wears the metal anchor insignia of all French Colonial Troops.*

mental distinction, were in white metal, or silver for Officers, and bi-metal (that is, a combination of brass and white metal). Some of the rules governing British (and other) badges had to be broken during World War I, because of the huge increase in number of men under arms and the need for metal for more vital purposes. Some wartime badges also were not made to the exacting standards of peacetime; notably, areas which traditionally were voided (ie, cut out) were left as solid metal for ease of manufacture. The large increase in size of the British Territorial Army in World War I brought with it a very large number of badge variations: many Territorial badges (*pages 54-55*) were based on those of the regular parent battalions, but with omissions or additions, creating a field large enough to tax even the most ardent collector. In some other countries, including France, metal cap badges were not issued; while in the German Army the cockades of State and Country on the headgear and the complex system of *Waffenfarbe* ("arm of service colours"), allied to metal numbers, on shoulder boards and collar patches (*pages 130-135*) distinguished units.

In British service, what has been said of badges applies in a more limited way to buttons. On these, because of their size, a smaller part of the regimental tradition is displayed—in the period 1914-24, only on the buttons of Officers and Senior Warrant Officers, for regimental pattern buttons were not restored to the rank and file until 1924 (except for the Guards and Household Cavalry, who always wore the regimental pattern). In buttons as in badges there are numerous variations; from Other Ranks' patterns for tunics, shoulder straps and caps, to Officers' buttons for tunics, full dress and mess dress. Greatcoat buttons, mounted buttons (insignia in silver or gilt mounted on a flat or domed button), flat buttons, domed buttons and hollow-backed buttons of many types and sizes must be considered by the serious collector. Another complex area is shoulder titles: as will be seen from the photographic section, there are as many variations in these as in cap badges and buttons. Collecting metal shoulder titles is in itself a full time hobby.

In France, regimental buttons bore simple numbers. In Germany buttons bore either a unit's numeral on a plain ground, the crest of the State from which the unit originated, or a design taken from the arms of the monarch or nobleman who was the unit's honorary commanding officer. American uniform buttons tended to be purely of a "general service" pattern, but US collar discs (*pages 64-65*) display a far greater variation.

PLASTIC & "STAYBRITE"

During World War II, many British cap and collar badges (the latter generally being a smaller variation on the cap badge, repeating its major motif) were made in plastic as an economy measure. They appeared in three colours, chocolate brown, beige and grey, and most (but not all) infantry regiments and all corps received them. None were made for cavalry regiments. Generally disliked by soldiers, these badges were not so carefully preserved (and, of course, were not so durable) as metal badges; they are, therefore, now quite rare—and consequently expensive.

Another major change in badge material came about in 1952 when the British Army began to receive anodised aluminium, or "Staybrite", badges and buttons. Although these, unlike the metal types, do not need cleaning, they are generally unpopular both with servicemen and with collectors, who dislike their cheap-looking glitter effect. Very few are shown on the following photographic spreads.

BRITISH CAP BADGES

Probably the single most popular category of badges for collectors all over the world is that of British Army cap badges. The most eagerly sought after are those of the period from World War I to 1922; in the latter year, the British Army's Irish regiments were disbanded with the establishment of the Irish Free State and certain cavalry regiments were amalgamated.

Simply because of the numbers of men and units involved, there are more cap badges in existence from World War I than from any other period. In many cases, different battalions of the same regiment had different cap badges, as in the case of the London Regiment (see *pages 50-51*), whose badges ranged from the crowned harp of the 18th Battalion (London Irish) to the White Horse of Kent of the 20th Battalion (Blackheath and Woolwich).

Cavalry cap badges (see *pages 46-47*) are extremely popular with collectors, but because there were fewer of these units than infantry, prices are higher. Probably the most famous of these badges is that of the 17th/21st Lancers: a skull and crossbones in white metal above a scroll bearing the motto "Or Glory", which has given them the nickname of "The Death or Glory Boys".

The 17th Lancers (Duke of Cambridge's Own) took part in the Charge of the Light Brigade during the Crimean War; they were merged in 1922 with the 21st Lancers (Empress of India's). Six new cavalry regiments, their mounts tanks rather than horses, were raised in World War II but disbanded in 1948. These war-raised units were the 22nd Dragoons, 23rd Hussars, 24th Lancers, 25th Dragoons, 26th Hussars and 27th Lancers, and their badges are, obviously, somewhat scarcer than those of longer-serving units.

The cap badges of British infantry of the line form the backbone of many badge collections. The Army of 1914-18 and later was mainly made up of regiments associated with the counties of Britain; many have now been disbanded or have been amalgamated with other units to form new regiments (and, of course, new badges). But the badges of famous regiments no longer in existence, such as the King's Own Yorkshire Light Infantry, Royal Northumberland Fusiliers, Royal Norfolk Regiment, South Wales Borderers, Manchester Regiment and York and Lancaster Regiment, can still be found quite easily and are not too expensive. Many are shown on *pages 52-53, 54-55.*

World War II saw the formation of a number of specialist units, such as the Parachute Regiment, formed in 1942, whose cap badge is worn in conjunction with the famous maroon beret, the hallmark of the British airborne soldier. Infantry battalions assigned as glider-borne troops in World War II usually wore their own regimental badge on the "red beret". The Special Air Service (SAS), whose winged-dagger badge (see *Inset, pages 46-47*) with the motto "Who Dares Wins" is now world-famous, was formed in North Africa in 1941 by Major David Stirling and is now known as 22nd Special Air Service Regiment. Its Territorial Army unit, 21st SAS, bears the "Mars and Minerva" cap badge (see (30), *pages 50-51*) of the Artists Rifles of World War I. Many Commando units of World War II did not wear cap badges, but a few had them locally made in the Middle East (and No 2 Commando reputedly made their own white-metal dagger insignia from stolen spoons!) After the War, most of these units were either disbanded or merged into the SAS, Royal Marines or other regiments.

Brass shoulder titles are gaining in popularity in the collecting field, mainly because of the rising cost of cap badges. Titles can still be picked up comparatively cheaply and may form a self-contained collection or, as seen on our photographic spreads, part of a collection of badges and buttons. There are an enormous number of different types, ranging from self-explanatory regimental and corps titles—such as "RA" (Royal Artillery), "RM" (Royal Marines) and "RDG" (Royal Dragoon Guards)—to such complex variations as those worn by the Duke of Wellington's Regiment (West Riding). Up to 1915 the unit wore the metal title "W. Riding", which was then changed to "Duke of Wellington's" or "Duke of Wellington". In 1931 it became "DW" and in 1970 "DWR" (in anodised aluminium). The titles "D of W" and "D of W's" may also be found.

AMERICAN BADGES

The collecting of American military badges is rapidly gaining in popularity, although for some years items of US militaria were ignored by the collecting fraternity in Europe. In the United States, a number of societies cater for collectors of American items, although most of these organisations are for the "rag pickers", as the collectors of cloth patches are known. Most new collectors of US metal badges begin with the metal "wings" (*pages 60-61*) worn by the US Army Air Force, 1941-1945. These, unlike the insignia of the Royal Air Force, which did not have all-metal wings, are very attractive pieces. Some of them are in solid silver, but these command very high prices and are seldom seen for sale. When the US Air Force was separated from the US Army in 1947, US Army pilots were issued new "wings", more upswept than the new USAF types.

Many collectors are also attracted by US Airborne insignia, of which there are many examples, ranging from the simple Parachutist Wing to the enamelled Airborne Pathfinder badge. Special Forces serving in Vietnam had a number of unofficial badges and wings made, most incorporating in the design a grinning skull wearing a green beret. These tend to be rare and fairly expensive. Distinctive Insignias, the small, enamelled badges worn on the epaulette or beret and commonly known as "DIs", are very attractive items. There is plenty of scope for the collector: at the last count, there were more than 25,000 different types—not including variations. Most designs refer to the history of the unit. For example, if the regiment fought in the Civil War a blue or white St Andrew's Cross is shown, while service in the Indian Wars is represented by an arrow or bundle of arrows or an Indian tepee. The Spanish-American War is symbolised by a castle; and regiments involved in the Boxer Rebellion in China have a dragon on their DI. Service in France in World War I is normally symbolised by a fleur-de-lis. Most DIs can be purchased at reasonable prices, although those for Airborne units are more expensive.

An excellent collection may also be formed of US Branch of Service insignia (*pages 64-65*), the badges or "collar dogs" worn on uniform lapels to denote the soldier's military role. Displayed in a frame, perhaps with a US Army

Above: *US Navy Officer with cap badge of post-May 1941 pattern (the eagle faces to its right) and silver collar bars of a Lieutenant (Junior Grade).*

Left: *Women of the British Auxiliary Territorial Service (formed 1938; reformed 1949 as Women's Royal Army Corps); note the metal "ATS" shoulder titles; these, like the cap badge seen here, were the only pattern issued to the unit.*

cap badge at the top, they make an attractive show. Once again, these are not expensive, although pieces for such specialist branches as the Tank Destroyer Forces of World War II are quite hard to find. Generally speaking, American badges have not suffered too greatly at the hands of makers of reproduction items, although copies of some of the rarer USAF wings of World War II are starting to appear, as are some Special Forces badges reputedly made for units serving in Vietnam.

THIRD REICH BADGES

Nazi militaria has suffered for some years from dealers who offer reproduction badges and insignia for sale. The quality of these pieces varies from the excellent copies of such badges as the Panzer Assault, Luftwaffe Pilot and Luftwaffe Paratrooper, currently made in Vienna, to the crude "pot metal" copies turned out by the thousand in dealers' workshops all over the world. The flood of reproductions has had the unfortunate effect of making many collectors lose confidence in Nazi military badges, automatically assuming that any reasonably-priced item is faked. This has helped push up the prices of genuine items to a point beyond the reach of the average collector. Political considerations aside —and it is understandable that some people will find Nazi material distasteful—Nazi badges are generally well made and attractive, and form a valid part of modern military badge collecting. Badges like the Infantry Assault Badge, the Panzer Assault Badge and the General Assault Badge (all shown on *pages 66-67*) can still be quite easily found, as can Nazi Party badges, Hitler Youth insignia and the metal eagles worn on the uniform cap by the Army, Navy and Air Force (*Luftwaffe*). More rare are the various *Luftwaffe* qualification badges, such as the Pilot and Pilot/Observer badges. The badges awarded to the *Luftwaffe*'s Hermann Göring Division, which fought as infantry, included the Luftwaffe Ground Combat Badge and the Luftwaffe Panzer Badge and these are now almost never seen outside museums or in the displays of long-established collectors.

However, a collection of the Nazi Party "day badges" can be made at a reasonable cost. These were the small, stamped-metal badges sold in aid of Nazi Party funds at rallies. There are lterally hundreds of different types which will form a basis for a collection of Nazi militaria.

The badges and insignia worn by paratroops and airborne forces are rapidly growing in popularity. The paratrooper wings and insignia of all nations are keenly sought after, but bargains can still be found. French "Paras" badges are still widely available ranging from the simple but striking Foreign Legion Para beret badge (see *Inset, pages 96-97*) to the enamelled breast badges worn by the various paratroop regiments. An excellent collection of the finely-made "DIs" of the French Foreign Legion is shown on *pages 62-63* and *64-65*.

The Vietnam War is of interest to many collectors, some of whom specialise in the badges and insignia of the former Republic of South Vietnam. Others concentrate on the badges worn by the various nations involved in that war alongside US Forces, such as South Korea, the Philippine Republic and Australia. Bargains can still be found, although prices are rising fast at the time of writing. There are not many collectors of military badges of the Communist bloc, mainly because of the scarcity of items for sale: a few pieces filter on to the market but, mainly because of a lack of reference material, there is little demand. These badges invariably incorporate a red star or a hammer and sickle in the design and vary in quality from excellent enamel work to poorly-made stamped tin plate.

FORMING A COLLECTION

Whatever type of collection is aimed for, the first requirement is good reference material. Without this, no proper theme or display can be planned and the collection deteriorates into a miscellaneous conglomeration of items put together without knowledge or thought. There are many excellent reference works on the market, covering all aspects of metal badge collecting. They range in price from inexpensive paperbound books on US badges to multi-volume studies of British regimental insignia. Many are listed in the "Bibliography" given on *pages 206-207*, where an effort has been made to include titles that should be readily available.

The first question to be answered is whether or not to specialise in any particular field. For the collector of British badges, an excellent start could be made by concentrating on one particular regiment or on a single brigade, collecting the badges, buttons and titles of all its sub-units, including the volunteer battalions of World War I. Alternatively, one may decide to collect cavalry or infantry badges, or badges from a particular county. If the latter course is decided upon there is great scope: for example, a specialist in Yorkshire regiments would need the metal insignia of at least six line regiments and three Yeomanry regiments, as well as modern amalgamations.

There are several different methods of displaying a collection; the reader must decide which is the most suitable. Some collectors mount badges in large frames to hang on a wall. Others prefer to mount them on cards, which are then inserted into plastic-pocket binders. Whichever method is selected, the badges should be cleaned before mounting—*not* with an abrasive metal polish, but with warm soapy water, using a toothbrush to get dirt out of the crevices. They should then be carefully dried before being mounted.

Reproductions of Nazi badges have already been mentioned, but the warning must be repeated in connection with the universally popular British badges. There are a number of unscrupulous dealers who are having certain rare badges made, sometimes from the original dies, to be passed off to the unsuspecting collector as genuine items —at grossly inflated prices. The most commonly reproduced badges are those of pre-1922 cavalry regiments, territorial battalions of World War I and elite forces of World War II, such as the Commandos, Long Range Desert Group and "Popski's Private Army". "Restrikes", as they are commonly known, are sometimes easily spotted because the metal used is lighter than that of the originals, or, in some cases, especially those badges emanating from Pakistan, heavier, because these are castings rather than die-struck. The only certain way to avoid buying restrikes is to purchase from a dealer who is willing to guarantee the authenticity of any badges he sells—although even dealers can make mistakes.

Metal Insignia of the British Cavalry

1-3 Cap Badge; Button, *c.* 1902-59; Shoulder Title; The Queen's Bays (2nd Dragoon Guards).
4-6 Cap Badge 1915-37; Button, 1855-1959; Shoulder Title; 1st King's Dragoon Guards.
7-9 Cap Badge, pre-1935; Button, 1855-1922; Shoulder Title, post-1920; The Inniskillings (6th Dragoons/5th Inniskilling DG).
10-12 Cap Badge, pre-1922; Button, 1855-1928; Shoulder Title; 3rd Dragoon Guards.
13-14 Cap Badge, post-1922; Button; 3rd Carabiniers (Prince of Wales's Dragoon Guards).
15 Shoulder Title, 1922-28; 3rd/6th Dragoon Guards.
16-17 Cap Badge, pre-1922; Button, 1855-1904; 4th Royal Irish Dragoon Guards.
18-19 Cap Badge, post-1935; Shoulder Title, post-1935; 5th Royal Inniskilling Dragoon Guards.
20-21 Cap Badge & Shoulder Title, 1907-58 (when amalgamated); 4th Queen's Own Hussars.
22-23 Cap Badge, 1898-1915; Shoulder Title, 1913-21; 1st King's Dragoon Guards.
24-25 Cap Badge, 1902-36; Shoulder Title, post-1921; The Royal Dragoons (1st Dragoons).
26-28 Cap Badge, post-1922; Button; Shoulder Title, post-1936; 4th/7th Royal Dragoon Guards.
29-32 Cap Badge; Collar Badge; Button; Shoulder Title, post-1920; The Royal Scots Greys.
33-34 Cap Badge, 1902-22; Shoulder Title, pre-1920; 6th Dragoon Guards (Carabiniers).
35-36 Cap Badge; Shoulder Title; 3rd The King's Own Hussars.
37-38 Cap Badge & Shoulder Title, pre-1922; 7th Dragoon Guards.
39 Cap Badge (OR; Officer in bronze), 1940-48; 22nd Dragoons.
40-41 Cap Badge & Shoulder Title, pre-1969; 10th Royal Hussars.
42-43 Cap Badge & Shoulder Title, pre-1969; 11th Hussars.
44 Field Service Cap Badge, *c.* 1900; 13th Hussars.
45 Cap Badge, 1902-22; 5th Dragoon Guards.
46-49 Cap Badge (OR), pre-1958; Button; Shoulder Title (OR; should read '8H'); Shoulder Title (Officer or RSM); 8th King's Royal Irish Hussars (amalgamated 1958).
50-52 Cap Badge, 1915-29; Cap Badge, pre-1915; Shoulder Title (OR); 14th King's Hussars.
53 Cap Badge (OR; Officer in bronze), 1941-48; 25th Dragoons.
54 Cap Badge, 1901-22; 13th Hussars.
55-57 Cap Badge, Button & Shoulder Title (OR); 1897-1922; 21st Lancers (Empress of India's).
58-60 Collar Badge, Cap Badge & Shoulder Title (Officer); pre-1922; 15th The King's Hussars.
61 Cap Badge (trial type), 1941-48; 26th Hussars.
62 Button, 1922 on; 16th/5th Lancers.
63 Cap Badge, 1911-22; 18th Royal Hussars (Queen Mary's Own).
64-65 Cap Badge, 1902-09; Cap Badge 1898-1902; 19th Royal Hussars (Queen Alexandra's Own).
66 Cap Badge, pre-1922; 16th The Queen's Lancers.
67 Shoulder Title, post-1922; 16th/5th Lancers.
68-71 Cap Badge, Button ,Shoulder Title (issue pattern), Shoulder Title (regimental pattern); pre-1960;

Special Air Service Regiment (SAS) cap badge, with the motto "Who Dares Wins", on khaki beret. Also shown are the cloth badge for wear on the beret; shoulder title; and qualified parachutist's badge.

9th Queen's Royal Lancers.
72 Cap Badge, Army Remounts Service (1914-1918).
73 Cap Badge (OR; Officer, bronze), 1940-48; 26th Lancers.
74 Cap Badge (OR; Officer, bronze), 1941-48; 27th Lancers.
75-76 Cap Badge (Officer; OR, white metal) & Shoulder Title (Officer); 1941-48; 26th Hussars.
77-78 Cap Badge (OR) & Shoulder Title; 1940-48; 23rd Hussars.
79-81 Cap Badge; Button & Shoulder Title (OR, regiment's favoured pattern), pre-1960; 12th Royal Lancers (Prince of Wales's).
82-83 Cap Badge, post-1922; Shoulder Title, 1922-34; 15th/19th The King's Royal Hussars.
84-85 Cap Badge & Shoulder Title, pre-1922; 20th Hussars.
86-88 Cap Badge, 1922-36, & Shoulder Title, 1922 on; 14th/20th Hussars; Collar Badge, pre-1922; 14th King's Hussars.
89-90 Cap Badge & Shoulder Title, post-1922; 17th/21st Lancers.
91-93 Collar Badge, Button & Shoulder Title, pre-1922; 17th Lancers.
94-95 Cap Badge & Button, 1917-22; Tank Corps ("Royal" in 1923).
96 Shoulder Title, 1933-39; Royal Tank Corps ("Regiment", 1939).
97 Cap Badge, 1922 on; Royal Tank Corps/Royal Tank Regiment.
98-99 Button & Shoulder Title, 1939 on; Royal Tank Regiment.
100-101 Cap Badge, 1939-41; Shoulder Title, 1939 on; Royal Armoured Corps (formed 1939).
102-103 Cap Badge & Button, 1941 on; Royal Armoured Corps.
104-105 Cap Badge & Button, 1941-1946; Reconnaissance Corps.

Metal Insignia: Guards, Royal Artillery, Yeomanry

1-3 Cap Badge (ranks below full sergeant); Button (King George V-VI pattern); Shoulder Title (letters surmounted by separate grenade to 1920, thereafter by badge bearing the Royal Cypher); Grenadier Guards.
4-6 Cap Badge (ranks below WO; higher ranks have smaller, silver badge with cross in red enamel); Officer's shoulder rank insignia; Button, 1855 on; Coldstream Guards.
7-9 Cap Badge (ranks below full sergeant); Button (King George V-VI pattern); Shoulder Title (letters surmounted by thistle badge in brass to be worn on service dress); Scots Guards.
10-11 Shoulder Title (service dress; leek alone worn on full dress until 1939); Puggaree Badge (worn on the cloth band around the tropical helmet); Welsh Guards.
12-13 Cap Badge (ranks below WO); Button (King George V-VI pattern); Irish Guards.
14-15 Shoulder Title (star worn above letters with service dress; letters not worn on full dress); Irish Guards.
16 Buttons, 1873 on; Royal Horse Artillery.
17 Button; post-1924; Royal Artillery.
18 Shoulder Title (service dress); Royal Horse Artillery.
19 Side Cap Badge, 1902 on; Royal Artillery.
20 Cap Badge; King George VI period on; Royal Horse Artillery.
21 Sleeve Badge (worn above chevrons by senior sergeants); Royal Artillery.
22 Shoulder Title (service dress): Royal Field Artillery.
23 Shoulder Title (service dress); Royal Garrison Artillery.
24-25 Cap Badge; Shoulder Title (service dress); Royal Artillery.
26 Cap Badge, 1918-20; Guards Machine Gun Regiment (worn only by 4th Battalion (Foot Guards); other three Battalions (Household Cavalry) retained own cap badges and wore Machine Gun Regiment collar badges).
27-29 Buttons; Cap Badge; Shoulder Title; 1915-22; Machine Gun Corps.
30 Cap Badge, 1914-19; Army Cyclist Corps.
31 Cap Badge, 1917-18; Women's Army Auxiliary Corps (redesignated Queen Mary's Army Auxiliary Corps, 1918; disbanded 1921).
32 Cap Badge; 1942-50; Army Air Corps.
33 Cap Badge, 1950-57; Glider Pilot Regiment.
34 Cap Badge, pre-1949; Queen Alexandra's Imperial Military Nursing Service.
35 Buttons, 1949 on; Queen Alexandra's Royal Army Nursing Corps.
36 Sleeve Badge; Marksman.
37-38 Button & Cap Badge, 1943 on; The Parachute Regiment.
39 Sleeve Badge; Drummer.
40 Sleeve Badge; Central Band, Royal Air Force; Army Bandsmen (except Cavalry) wore the same badge but without the crown.
41 Cap Badge, World War II period; No 2 Commando.
42 Cap Badge, 1947 on; Special Air Service Regiment.
43 Sleeve Badge; Artificer.
44 Cap Badge; Duke of Lancaster's Own Yeomanry.

45 Cap Badge; Cheshire Yeomanry (Earl of Chester's).
46 Cap Badge, pre-1953; Lanarkshire Yeomanry.
47 Cap Badge, Montgomeryshire Yeomanry.
48-49 Buttons & Cap Badge, pre-1953; The Queen's Own Worcestershire Hussars.
50-51 Cap Badge (OR; Officer, knot enclosed in Garter and motto with scroll below); Shoulder Title; Staffordshire Yeomanry.
52-53 Cap Badge & Shoulder Title pre-1956; The Fife and Forfar Yeomanry.

Dress uniform tunic of a Major, Royal Engineers. Note that collar badges resemble those of Royal Artillery, but with nine flames from grenade rather than seven.

54-55 Cap Badge; Shoulder Title, 1952 on; Sherwood Rangers Yeomanry.
56-57 Cap Badge, pre-1936; Shoulder Title; North Somerset and Bristol Yeomanry.
58-59 Cap Badge (worn above scroll with regimental title); Shoulder Title; Denbighshire Yeomanry (Hussars).
60 Cap Badge; Pembroke Yeomanry.
61 Cap Badge (OR; Officer, silver and gilt); 1923-66; The Ayrshire Yeomanry (Earl of Carrick's Own).
62-63 Cap Badge & Shoulder Title; pre-1953; The Hampshire Carabiniers Yeomanry.
64-65 Cap Badge (note special pattern crown both here and at (64)); Shoulder Title, post-1950; Royal Gloucestershire Hussars.
66 Cap Badge; pre-1950; Westmorland and Cumberland Yeomanry.
67 Cap Badge (OR), 1908-14; The Lothians and Border Horse Yeomanry.
68 Cap Badge, pre-1961; The Berkshire Yeomanry.
69 Cap Badge; The Northumberland Hussars.
70 Cap Badge; West Somerset Yeomanry.
71 Cap Badge; Leicestershire Yeomanry (Prince Albert's Own).
72 Cap Badge; Glamorgan Yeomanry.
73 Cap Badge; 1901-20 (when disbanded); Lincolnshire Yeomanry.
74 Cap Badge; Queen's Own Oxfordshire Hussars.
75 Cap Badge; Lancashire Hussars.
76 Cap Badge; The Duke of York's Own Loyal Suffolk Hussars.
77 Cap Badge, pre-1953; The Derbyshire Yeomanry.

1-6 Shoulder Title, Collar Badge, 1923 on; Cap Badge, 1923 on with 1952 change in crown; Buttons (full dress); Belt Buckle, for white buff belt; Collar Badge (Officer); Royal Marines.
7 Collar Badge, 1917-23; Royal Marine Artillery.
8 Cap Badge, Royal Marine Labour Corps; formed February 1917 and disbanded 1919.
9 Button, 1871-1924; General Service.
10-11 Cap Badge, Button, pre-1923; Royal Marine Artillery.
12 Shoulder Title, Royal Marine Band.
13-14 Cap Badge, Collar/Sleeve Badge; Royal Marine Bandsman, Royal Naval School of Music.
15 Shoulder Title, 1948-64; Royal Marine Forces Volunteer Reserve.
16 Cap Badge, 1923 on; Royal Marines.
17-19 Shoulder Title, Cap Badge, Buttons, 1898-1923; Royal Marine Light Infantry.
20-23 Collar Badge, Button, Shoulder Title, Cap Badge, post-1962; Royal Marines.
24 Cap Badge, pre-1937; 12th (County of London) Battalion, The London Regiment (The Rangers).
25 Cap Badge, pre-1937; 17th (County of London) Battalion, The London Regiment (Poplar and Stepney Rifles).
26 Cap Badge, pre-1937; 18th (County of London) Bn., The London Regiment (London Irish Rifles).

Other Rank's Glengarry, with post-1908 badge of 14th (County of London) Battalion, The London Regiment (London Scottish).

27 Cap badge, pre-1921; 8th (City of London) Battalion, The London Regiment (Post Office Rifles).
28 Cap Badge, pre-1920; 25th (County of London) (Cyclist) Battalion, The London Regiment.
29 Cap Badge, 1902 on (with change of crown); The Honourable Artillery Company.
30-33 Cap Badge, Button, Shoulder Title, Button, pre-1937; 28th (County of London) Battalion, The London Regiment (Artists Rifles); since 1947, the badge has been worn by 21st Bn., Special Air Service Regiment (Artists Rifles) (T.A.).
34-38 Cap Badge, 1908-20; Cap Badge, 1920-56; Button (horn); Shoulder Title, 1908; Button (bronzed); 5th (City of London) Battalion, The London Regiment (London Rifle Brigade); from 1937, The London Rifle Brigade, The Rifle Brigade.
39-42 Cap Badge, Button, Shoulder Title, Button, pre-1935; 19th (County of London) Battalion, The London Regiment (St Pancras).
43-44 Cap Badge, post-1908; Shoulder Title, post-1915; 14th (County of London) Bn., The London Regt. (London Scottish).
45-47 Cap Badge, Collar Badge, Shoulder Title, pre-1918; Army Ordnance Corps.
48-52 Collar Badge, Button, Cap Badge, Shoulder Title, Button, post-1918; Royal Army Ordnance Corps.
53-58 Cap Badge, Collar Badge, 1942-46; Cap Badge, Button, Collar Badge, Shoulder Title, 1946-53; Royal Electrical and Mechanical Engineers.
59 Cap Badge, 1920-46; Army Educational Corps.
60 Cap Badge, 1906-18; Army Veterinary Corps.
61 Cap Badge, 1940-56; Intelligence Corps.
62 Cap Badge, 1938-49; Auxiliary Territorial Service (ATS).
63 Button, post-1949; Women's Royal Army Corps.
64 Shoulder Title, 1938-49; ATS.
65 Cap Badge, 1949-53; Women's Royal Army Corps.
66 Cap Badge, 1940-52; Army Physical Training Corps.
67-68 Cap Badge, Button, 1902-26; Military Foot Police.
69-70 Cap Badge, Button, 1926-46; Corps of Military Police.
71-73 Cap Badge, Buttons, Shoulder Title, 1902-46; Royal Army Medical Corps (RAMC).
74-76 Cap Badge, 1920-47; Buttons; Shoulder Title, pre-1929; Royal Corps of Signals.
77-80 Cap Badge (plastic), 1942-44; Cap Badge, 1953; Shoulder Title, Button, pre-1946; Pioneer Corps (from 1946, Royal Pioneer Corps)
81-84 Cap Badge (silver, denoting Territorial volunteer), 1898 on; Cap Badge, 1914-18; Shoulder Title, 1910; Shoulder Title (Signal Service, Territorial), pre-1920; Royal Engineers.
85-89 Cap Badge, 1902-18; Cap Badge, post-1952; Collar Badge, Shoulder Title, 1902-18; Shoulder Title, post-1918; Button, 1937-53; (Royal) Army Service Corps.
90-93 Cap Badge, 1929-55; Forage Cap Badge, pre-1902; Shoulder Title, 1929-46; Collar Badge, 1946 on; (Royal) Army Pay Corps.

Metal Insignia of the British Infantry

1-4 Cap Badge with red cloth backing, 1902 on; Shoulder Title, post-1921; Button, 1881 on (Officer), 1924 on (OR); Shoulder Title, pre-1921; The Royal Scots. Like other cap badges and buttons shown here and on other spreads in this section, the cap badge will also be found in anodised aluminium ("Staybrite") form, dating from 1952 onward. These permanently bright badges, needing no cleaning, are generally unpopular with servicemen and collectors alike, and some units are known to have ordered privately-made badges in the original metals.

5-7 Cap Badge, pre-1926; Button, 1881 on (Officer, Senior NCO), 1926 on (OR); Shoulder Title, post-1908; The King's (Liverpool Regiment), amalgamated 1958.

8-10 Cap Badge, 1902-52; Collar Badge, 1902 on; Shoulder Title, post-1908; The Royal Fusiliers (City of London Regiment).

11-13 Cap Badge, 1898-1961; Button, 1881 on (Officer), 1924 on (OR); Shoulder Title, post-1902; The Buffs (Royal East Kent Regiment).

14-16 Cap Badge, pre-1920; Button, 1909 on (Officer), 1924 on (OR); Shoulder Title, post-1902; The Queen's (Royal West Surrey Regiment), amalgamated 1959.

17-19 Cap Badge, pre-1946; Button, 1881-1946 (Officer), 1924 on (OR); Shoulder Title, post-1902; The Lincolnshire Regiment (Royal Lincolnshire Regiment from 1946).

20-22 Cap Badge, 1898-1950; Button, post-1881 (Officer), 1924 on (OR); Shoulder Title, 1918-20; The Prince Albert's (Somerset Light Infantry).

23-25 Cap Badge, 1898-1935; Button, 1881-1935 (Officer), 1924 on (OR); Shoulder Title, 1902-35; The Norfolk Regiment.

26-27 Cap Badge, 1898-1958; Button, 1881-1958 (Officer), 1924 on (OR); The Prince of Wales's Own (West Yorkshire Regiment).

28 Shoulder Title, pre-1920; The Princess of Wales's Own (Yorkshire) Regiment.

29 Shoulder Title, post-1920; The Prince of Wales's Own (West Yorkshire Regiment).

30-32 Cap Badge, Button, 1919-58; Shoulder Title, 1919; The Bedfordshire & Hertfordshire Regiment.

33-35 Cap Badge, 1898-1946; Button, 1881-1960; Shoulder Title, pre-1921; The Leicestershire Regiment (Royal Leicestershire Regiment from 1946; button not altered).

36-38 Cap Badge; 1905-51; Button, 1903-51; Shoulder Title, post-1920; The Princess of Wales's Own (Yorkshire) Regiment (after 1920, The Green Howards (Princess of Wales's Own Yorkshire Regiment).

39-42 Glengarry Badge; Grenade Badge (worn above shoulder title and on collar with full dress); Collar Badge (Officer, undress); Shoulder Title; 1901-02 on; The Royal Scots Fusiliers.

43-45 Cap Badge, 1901-56; Button, 1883-1958; Shoulder Title, 1902; The Devonshire Regiment.

46-48 Cap Badge, 1898-1922; Button, 1881 on; Shoulder Title, 1902; The Cheshire Regiment.

49-51 Glengarry Badge, 1901-52; Button, 1901 on; Shoulder Title, 1902 onward; The King's Own Scottish Borderers.

52-54 Cap Badge, 1898-1958; Button, 1887 on; Shoulder Title, 1902; The South Wales Borderers.
55-58 Cap Badge, 1898-1934 (flag flying to left; after 1934 to right); Button, 1881 on; Shoulder Title, pre-1921; Collar Badge; The Royal Inniskilling Fusiliers.
59-64 Collar Badge; Cap Badge, 1898-1958; Shoulder Titles, pre-1931, 1920-31, pre-1915, 1915-31; The Duke of Wellington's Regiment (West Riding).
65-67 Bonnet Badge, 1898-1968; Shoulder Titles, pre-1921, post-1921; The Cameronians (Scottish Rifles), disbanded 1968.
68-70 Cap Badges (front, rear), pre-1958; Shoulder Title, 1902; The Gloucestershire Regiment.
71-73 Cap Badge, 1898-1956; Shoulder Titles, pre-1914, post-1914; The Duke of Cornwall's Light Infantry, amalgamated 1959.
74 Cap Badge (Officer), 1898-1952; The Border Regiment.
75 Collar Badge, 1902-46; The Hampshire Regiment (from 1946, The Royal Hampshire Regiment).
76-78 Collar Badge; Cap Badge, 1898-1952; Shoulder Title, post-1902; The Dorsetshire Regiment, amalgamated in 1958.
79-81 Cap Badge, 1898-1958; Button, 1885-1958; Shoulder Title, 1902; The South Staffordshire Regiment.
82-84 Cap Badge, 1898-1958; Button, 1885-1958; Shoulder Title, 1902; The Royal Sussex Regiment.
85-88 Cap Badge, pre-1946 (when "Royal" added); Button, 1881; Shoulder Titles, pre-1921, post-1921; The Hampshire Regiment.
89 Cap Badge, 1916 (wartime issue, all brass; peacetime issues bi-metal); The Prince of Wales's Volunteers (South Lancashire Regt).
90-96 Cap Badge, Shoulder Title, Collar Badge, pre-1920; Button, 1920-60; Mess Dress Button; Cap Badge, post-1920 (with new spelling); Shoulder Title, 1920 (new spelling); The Welch Regiment (until 1920, when spelling changed, The Welsh Regiment).

Service Dress jacket, with Sam Browne belt, of a Captain, Royal Warwickshire Regiment. The collar badges are of the same design as the Regiment's cap badge. The jacket, badges and buttons date from World War II; but note that the shoulder "pips" signifying rank are anodised aluminium ("Staybrite") and are thus of a later date.

Metal Insignia: British Infantry and Territorials

1-4 Glengarry Badge (Sergeant), pre-1926; Glengarry Badge, pre-1934; Button, pre-1934; Shoulder Title, 1921-27; The Black Watch (Royal Highland Regiment).

5-8 Cap Badge, pre-1945 (when smaller version issued for beret); Shoulder Title, 1958 on (after amalgamation into The Green Jackets Brigade); Button, 1908-46; Shoulder Title, 1908-18 (worn with bugle horn badge above it); The Oxfordshire and Buckinghamshire Light Infantry.

9-12 Cap Badge, 1898-1959; Button, 1902-59; Shoulder Title, pre-1885 (when "Royal" granted); Shoulder Title, c1907-20; The Royal Berkshire Regiment (Princess Charlotte of Wales's).

13-15 Cap Badge, 1898-1960; Button; Shoulder Title, 1908-20; The Northamptonshire Regiment.

16-18 Cap Badge, 1902-70 (change of crown in 1952); Shoulder Title, 1902-20; Shoulder Title, post-1920; The Sherwood Foresters (Nottinghamshire and Derbyshire Regiment).

19-21 Cap Badge, 1898-1961; Button; Shoulder Title, 1908-39; The Queen's Own Royal West Kent Regiment.

22-24 Cap Badge, 1898-1966; Button; Shoulder Title, 1908-39; The Middlesex Regiment (Duke of Cambridge's Own).

25-27 Cap Badge, 1905-52 (when plastic badge with Queen's crown issued; red cloth backing is standard); Button; Shoulder Title, pre-1958 (when amalgamated into The Green Jackets Brigade); The King's Royal Rifle Corps.

28-29 Cap Badge, 1898-1959; Shoulder Title, 1902 on; The Wiltshire Regiment.

30-32 Cap Badge, 1898-1959; Button; Shoulder Title, post-1902; The North Staffordshire Regiment (The Prince of Wales's).

33-34 Cap Badge, 1898-1968; Shoulder Title, 1902 on; The York and Lancaster Regiment.

35-36 Cap Badge, 1901-59; Shoulder Title, post-1930; The Highland Light Infantry (City of Glasgow Regiment).

37-39 Cap Badge, pre-1923; Button; Shoulder Title, 1908 on; The Manchester Regiment.

40-43 Cap Badge, 1898-1958; Collar Badge, 1902 on; Button; Shoulder Title, 1902 on; Seaforth Highlanders (Ross-shire Buffs, The Duke of Albany's).

44-47 Cap Badge, 1898 on; Collar Badge; Button; Shoulder Title, 1902-22 (when changed to "GORDONS"); The Gordon Highlanders.

48-51 Cap Badge, 1898 (or c1912)-1961; Button; Shoulder Title, post-1924; Collar Badge, 1902 on; The Queen's Own Cameron Highlanders.

52-54 Cap Badge, 1898-1922; Button, pre-1922; Shoulder Title, c1914-22; The Connaught Rangers.

55 Cap Badge, 1913 (previously in black)-1958 (with change of crown); The Royal Irish Rifles (from 1920, The Royal Ulster Rifles.

56 Shoulder Title, post1920; The Royal Ulster Rifles.

57 Shoulder Title, pre-1920; The Royal Irish Rifles (which became the Royal Ulster Rifles in 1920.

58-60 Cap Badge, 1898 on; Collar Badge, 1902 on; Shoulder Title, 1902 on; The Argyll and Sutherland Highlanders (Princess Louise's).

61 Cap Badge, 1898-1922 (when disbanded); The Royal Munster Fusiliers.

62-63 Cap Badge, 1898-1922 (when disbanded); Collar Badge; The Royal Dublin Fusiliers.

Balmoral Bonnet with white-metal badge (1898-1958) worn by an Other Rank (Warrant Officers had the stag's head only, in silver plate) of the Seaforth Highlanders (Ross-shire Buffs, The Duke of Albany's). Also shown is an Officer's Sporran. In 1961 the Regiment was amalgamated with The Queen's Own Cameron Highlanders and then formed The Queen's Own Highlanders (Seaforth and Camerons).

64-66 Cap Badge, 1898-1922 (when disbanded); Button, pre-1922; Shoulder Title, c1914-22; The Prince of Wales's Leinster Regiment (Royal Canadians).

67 Cap Badge, 1910-37; The Rifle Brigade.

68 Cap Badge, January 1942-1943 The Highland Regiment.

69 Cap Badge, January 1942-1943 The Lowland Regiment.

70 Cap Badge, 1914-18; 11th (Lonsdale) Battalion, The Border Regiment. Volunteer battalion raised by Lord Lonsdale, who presented silver badges. Most were lost in an altercation with another unit during training, and brass badges were subsequently issued to replace them.

71 Cap Badge, 1908-18; 8th (Irish) Battalion, The King's (Liverpool Regiment).

72 Cap Badge, 1908 on; The Suffolk Regiment (Territorial Battalions).

73 Cap Badge, pre-1908; 3rd (Militia) Battalion, The Somerset Light Infantry (Prince Albert's).

74-75 Cap Badge, 1908-47; Shoulder Title, 1908-38; 6th (Duke of Connaught's Own) Battalion (TF), The Hampshire Regiment.

76-77 Cap Badge, Shoulder Title, 1908-43; 5th (Cinque Ports) Battalion (TF), The Royal Sussex Regiment.

78 Cap Badge, 1908 on; 4th, 5th and 6th Battalions, The Leicestershire Regiment.

79 Cap Badge, 1908-37; 10th (Scottish) Battalion (TF), The King's (Liverpool Regiment).

"Sweetheart Brooches": Britain and Commonwealth

"Sweetheart Brooches" were produced commercially and by servicemen as gifts for loved ones.

1 A collar badge ("collar dog") of the Royal Fusiliers, as a brooch.

2 A copy of the cap badge of the Cambridgeshire Regiment, with coloured enamel inserts.

3 The emblem of the Merchant Navy within a coiled rope motif.

4 A commercial design: cap badge of the Royal Fusiliers in silver.

5 Crest of the Army Ordnance Corps, a commercial design, enamelled.

6 An unusual brooch: Cinque Ports Battalion, Royal Sussex Regiment.

7 The Leicestershire Regiment.

8-11 Four variations on a popular theme of World War I: the flags of Allied nations—Great Britain, France, Belgium, Italy.

12 Collar Badge of the Scots Guards converted to a brooch (and shown here at the wrong angle).

13 Silver and enamelled brooch of the Lincolnshire Regiment.

14 Another variation on (8-11).

15 One of the most popular commercial products: silver insignia of the Royal Navy.

16 A commercial copy of the cap badge, Bedfordshire Regiment.

17 An odd variation of the Allied flags theme: the flags of Great Britain, Belgium, Italy, France and Japan are shown as five playing cards; hence the motto "NAP"—five tricks being the winning hand in that card game.

18 Brooch of the Royal Naval Air Service, a unit in existence only until 1918.

19 This Royal Naval Air Service brooch was converted from a badge.

20 Converted collar badge of the West Yorkshire Regiment.

21 The crest of 84 Squadron, Royal Air Force—typical of many such RAF Squadron brooches.

22 Royal Armoured Corps brooch.

23 Australia: not a badge brooch, but a typical commercial product.

24 Copy of the cap badge of The Queen's Royal Regiment (West Surrey).

25 Brooch in gilt and enamel of the Sussex Yeomanry.

26 Silver gilt brooch of the Royal Engineers; a converted collar badge.

27 An ornate and commercial Royal Navy brooch.

28 United States Forces in England: an all-purpose silver and enamel brooch of World War II.

29 Brooch of the HAPAG shipping line.

30 Observers Wings in silver, worn with full dress by RFC and RAF officers; not really a brooch, although perhaps worn as one.

31 Royal Artillery brooches, like this one in silver and enamel, are among the most common, since the RA formed about one-third of the British Army in both World Wars.

32 Royal Air Force: an early commercial example from the shape of the propeller.

33 66th Canadian Infantry Battalion: direct copy from the shoulder flash.

34-37 World War I brooches from the badges of Canadian units: 82nd Canadian Infantry Battalion; 81st CIB; 29th CIB; 106th CIB.

38-39 Commercially produced badges in gilt and enamel for the Machine Gun Corps of 1914-1919.

40 The Royal Marines emblem is

On its original card, this replica of the Northumberland Fusiliers' glengarry badge is one of a huge number of commercially produced "sweetheart brooches".

almost unrecognisable in this ornate silver and enamel commercial brooch.

41 Brooch for the Army Dental Corps.

42 All-purpose brooch in silver and enamel for Canadian forces.

43 An attractive silver and enamel brooch for the Women's Royal Naval Service (WRNS, or "Wrens").

44 A Lee-Enfield rifle replaces the mounting bar in this brooch for the Royal Sussex Regiment.

45 Wings brooch of the Royal Flying Corps in gilt and enamel.

46 Royal Artillery; gilt and enamel.

47 Royal Engineers; silver and enamel.

49 The "bomb" shows that this was a brooch for a Fusilier regiment.

49 Cap badge motif of the King's Royal Rifle Corps.

50 Cap badge motif, Cameron Highlanders.

51 Coldstream Guards: the oval badge denotes an officer.

52 Ornate Royal Navy brooch of obviously commercial origin.

53 Brooch featuring the cap badge of the Parachute Regiment.

54 RAF Pilot's Wings.

55 Brooch formed by mounting an epaulette button of the Royal Canadian Artillery on a stylised bar.

56 Royal Canadian Regiment: as (55).

57 Brooch of Royal Armoured Corps.

58 The crossed Vickers guns of the Machine Gun Corps.

59 Brooch featuring the cap badge of Queen Mary's Regiment (Surrey Yeomanry).

60 New Zealand Auxiliary Nursing Service: a uniform badge.

61 Royal Artillery: a silver badge worn on dress uniforms.

62 Silver and enamel brooch from a Royal Engineers' button.

63 Unusual Royal Flying Corps brooch; a hand-enamelled collar badge.

64 The emblem of The Buffs (Royal East Kent Regiment) is set within a "Good Luck" horseshoe.

65-72 British unit badges in "spur" or "wishbone" settings: (65) Royal Artillery; (66) Royal Electrical & Mechanical Engineers, original badge; (67) Royal Engineers, badge reproduced in regimental colours of red and blue; (68) Royal Pioneer Corps; (69) 4th & 5th Somerset Light Infantry, brooch converted from collar badge of the territorial unit; (70) Royal Army Service Corps, badge in World War II form; (71) Royal Army Medical Corps, a gilt and enamel brooch; (72) Royal Marines.

"Sweetheart Brooches": United States and Britain

1-3 Collar insignia, US Women's Army Corps (WAC): (1) Officer; (2) Enlisted Woman; (3) Officer— "subdued" for fatigue uniform.
4 1st Cavalry Division, US Army.
5 Lapel badge; Adjutant General's Department, US Army.
6 British "V-for-Victory" badge.
7 Lapel badge; British Prisoners of War Help Association.
8 Silver and enamel brooch with US coat of arms, World War I.
9 American eagle with enamel service flag, denoting that wearer had a sweetheart in the US forces.
10 Cape insignia; US Red Cross.
11 US Army Expeditionary Force; British-made brooch.
12 Gilt and plastic brooch; US arms.
13 Enamelled brooch: US shield with British and French flags and US Service flag suspended.
14 "Over There"; enamel Service flag.
15-16 Miniature US Army rank chevrons as brooches: (15) Master Sergeant; (16) Staff Sergeant.
17 US Honorable Discharge badge.
18 US Army Air Force (USAAF) squadron insignia brooch.
19 British-made gilt and enamel badge: USAAF insignia.
20 As (19); US Army Engineers.
21 Rare plastic badge commemorating attack on Pearl Harbor; c1942.
22 US arms in gilt, linked by chain to Field Artillery insignia.
23-24 Variations on (22), less chain.
25 Miniature enamel insignia of US 13th Airborne Division.
26 As (25); US 5th Army.
27 Lapel pin; US Quartermasters.
28 As (25); US Army Air Corps.
29 Lapel badge; US Navy Construction Battalions ("Seabees").
30 Gilt lapel badge; US Navy officer.
31 Lapel pin; US Army Signal Corps.
32 Enamel bracelet charm; US origin.
33 As (22); US Army Infantry.
34 Lapel pin; US Army Infantry.
35 Brass bracelet charm; US 9th AF.
36 Lapel badge; US Quartermasters.
37 Lapel badge; US Marine Corps Civilian Service.
38 Miniature rank insignia; US Army Colonel, worn as brooch.
39 "A Son in Service"; US parents.
40 United Service Organisation (USO) lapel badge; USO arranged entertainment for US troops overseas.
41 Lapel badge; US Red Cross.
42 US "V-for-Victory"; tin plate.
43 Recruit's badge, US Red Cross.
44 Junior US Red Cross badge.
45 Lapel pin (shown inverted); US Chemical Warfare Service.
46 Badge; US campaign hat with cannon insignia of Artillery.
47-48 Miniature US Army insignia: (47) Coastal Artillery; (48) Anti-Aircraft Artillery.
49 US arms linked by fine chain to insignia of 35th Infantry Div.
50 US arms suspended from USAAF insignia; both in silver.
51 Lapel badge; Marine Corps League.
52 Miniature Marine Corps insignia.
53 Variation on (52).
54 Variation in gilt on (49).
55-57 US Civil Air Patrol Achievement Ribbons: (55) "Charles Lindbergh"; (56) "General 'Hap' Arnold"; (57) "Rickenbacker".
58-61 US Service Ribbons, worn to denote a loved one in service: (58) Navy; (59) Navy Pilot; (60) Army Engineers; (61) Infantry.
62-63 Miniature US Campaign Medal Ribbons: (62) Asiatic-Pacific; (63)

(Top) USAF Pipe Band; unofficial Wing, 1950s; (Centre) US Balloon Pilot Wing, rare; (Bottom) "Sweetheart" miniature of US Airship Pilot Wing, silver brooch.

European-African-Middle East.
64-65 US Army Corps Aviator Wing in silver (64) and gilt (65).
66-67 US Army insignia: (66) Officer, Veterinary Corps; (67) Nurse Corps.
68-72 Miniature US Navy insignia: (68-70) Pilot; (71-72) Observer.
73 Miniature USAAF Pilot Wing.
74 USAAC insignia; gilt and enamel.
75 USAAF insignia; brass.
76 USAAF insignia with motto: "Keep 'em Flying"; civilian wear.
77-78 USAF Ground Observation Control badges, silver: (77) Supervisor; (78) Observer.
79 Miniature US Navy Flight Surgeon Wing; gold plate.
80-81 Miniature USAAF Wings: (80) Bombardier; (81) Senior Balloon Pilot—extremely rare.
82 Miniature Chinese Air Force Pilot Wing worn by members of American Volunteer Group ("Flying Tigers"), 1941-42.
83 Lapel badge; US 82nd Airborne.
84 Silver and enamel pin: US arms.
85 Lapel pin; USAAF.
86 Lapel badge; 10th AF, USAAF.
87-88 Miniature USAAF Wings: (87) Aircrew; (88) Observer.
89 US Army Air Corps identity photograph, mounted as brooch.
90-93 "V-for-Victory" brooches: (90) British-made; (91) US-made, USAAF insignia; (92) British-made, brass and enamel; (93) Plastic, with silver-wire "Sweetheart" and USAAF insignia.
94 B-17 "Flying Fortress"; silver.
95-96 Miniatures of US Army Expert Infantryman Badge.
97 Lapel badge; USA Service Clubs.
98 Miniature US Army Aviator Wing.
99 Miniature Wing featuring the Alamo; possibly from Texas.
100 Miniature USAAF Pilot Wing.
101 Miniature British Parachute Regiment badge.
102 British Parachute Qualification Wing, enamel.
103 Lapel pin; USAF Association.
104 Lapel badge; US 7th Air Force.
105 Silver charm bracelet; USAAF.
106 Necklet with 8th AF badge.
107 Brooch; Auxiliary Territorial Service (ATS); British cap badge suspended from forage cap.
108 Brass ATS badge with bow.
109 British Women's Royal Naval Service (WRNS) brooch; chromed.
110 ATS tie-pin in gilt and enamel.
111 Lapel badge; British Territorial Army (TA).
112 ATS tie-clip, brass.
113 British Spitfire badge.
114 Lapel badge; Canadian flag.

Metal Insignia: US Army, Navy and Air Force

1 US Army Air Defense School; breast-pocket badge worn only by students from foreign countries. Badges for a further 27 schools—each with a different enamelled centrepiece—are known.

2 US Army Recreation Service; breast-pocket badge worn by permanent staff and instructors.

3 US Army Tactical Combat Instructor; worn on right breast pocket by instructors at training centre, Fort Dix, N.J.

4 Rifle Sharpshooter, US Marine Corps; breast-pocket badge, worn below any medal ribbons, awarded to all Marines qualified as sharpshooters.

5 White House Service; authorised in 1961 for wear by service personnel on duty at the White House. A rare badge, with the Presidential Seal on a white-enamel gold-bordered disc.

6 Badge of Identification worn by members of the Honor Guard at the Tomb of the Unknown Soldier, Washington, D.C., and very rare.

7 US Public Health Service; officer's cap badge, World War II. Military officers on assignment wore this with their own uniforms.

8 US Navy, Surface Warfare. Gold badge, authorised 1976, worn on left breast pocket by officer qualified as Officer of the Deck or Junior Engineer Officer of the Watch.

9 44th Medical Brigade, Vietnam; worn on tab, right breast pocket, by members of the unit in Vietnam, 1966-70. Such metal "patches" are now rare.

10 US Army Glider badge; authorised 1944 for wear by personnel who made one or more combat drops with a glider-borne unit. This sterling silver miniature was for wear above medal miniatures on mess dress only.

11 Master Parachutist, 18th Airborne Corps; unofficial badge worn on jump suits by members of the unit's Free-Fall Para Team.

12 Master Parachutist, 82nd Airborne Division; as (11).

13 US Army Air Force Air Weather Service; unofficial badge for wear on civilian jacket by meteorology technician, World War II.

14 45th Air Ambulance Company, Vietnam; metal pocket patch—as (9)—worn 1967-71. Nickname and call-sign "Dust Off", forming part of the design of the badge derives from dust-storm created by helicopter's rotors.

15 US Army Transport Service; officer's cap badge, 1930-42.

16 1st Infantry Div; Vietnam, 1965-70, as (9).

17 2nd Field Force; Vietnam, 1965-70, as (9).

18 3rd Aviation Detachment (Assault Helicopter Company); Vietnam, 1964 and 1971-72, as (9), rare.

19 US Army Combat Medical badge; worn above any medal ribbons by medical personnel "who attain established standards or whose action in combat is exemplary".

20 US Air Force Combat Control Team; worn on fatigue jacket by men who parachute into forward combat areas to secure landing grounds and drop zones.

21 Airplane Observer Wings, US

US Army Air Force qualification insignia, "wings", World War II. Second row, right: Balloon Observer badge, 1919; reissued 1941 as Balloon Pilot badge.

Army Air Force; authorised 1921, now obsolete, for observers not qualified as pilots.

22-24 Chief (22) and Senior Flight Nurse (23), and Flight Nurse (24), US Air Force; new type wings authorised in 1977.

25 Senior Medical Specialist, US Air Force; from January 1978.

26 Chief Nurse, US Air Force; authorised January 1963.

27-29 Chief (27), Senior (28) and Air Crew Member; US Air Force wings for airmen who form part of flying crews; authorised 1942. See also (37).

30 Assistant Fire Chief, US Air Force; pocket badge.

31 Helicopter Crew Chief, US Army; unofficial wing worn by Army aviation crew chiefs, Vietnam, 1964-72; rare and much prized.

32 Command Pilot, US Army Air Force; authorised 1941 for senior pilots with long air time, several years of service and proven command abilities. See also (40).

33-35 Airman (33), Airman 1st Class (34) and Senior Airman (35), US Air Force; rank insignia, authorised 1976, worn on collar of topcoat or raincoat.

36 Junior Reserve Officer Training Corps (ROTC) Instructor, US Air Force; breast badge of the organisation giving basic Air Force training to the 13-17 years age group in high schools and junior colleges.

37 Air Crew Member, US Army Air Force; note variation from (29) in this sterling-silver badge—the later types shown at (27-29) are of oxidised 1/20th silver-filled metal and the feathering of the wings is less detailed.

38-39 Glider Pilot, US Army Air Force; authorised 1942, obsolete and now rare: (38) for wear on shirt; (39) regulation size.

40 Senior Pilot, US Army Air Force; authorised 1941 for pilot with specified flying hours, experience and time in grade.

41 Command Pilot Astronaut, US Air Force; authorised for pilots qualified on powered craft capable of flights more than 50 miles (80km) from Earth's surface, who have made at least one such flight.

42-44 Senior Pilot Astronaut (42), Pilot Astronaut, worn on shirt (43), and Pilot Astronaut, regulation size badge (44), US Air Force; qualifications as (41).

Metal Insignia of the French Foreign Legion

Regimental, unit and formation badges of the French Foreign Legion. Note that items marked with an (*) before dates of wear are re-strikes by the original maker, Drago et Bertrand, of virtually unobtainable badges. Produced with the agreement of the Legion and sold through its magazine *Képi Blanc*, these are generally acknowledged to be legitimate collectables.

1 Dépôt Commun de Légion Etrangère; worn 1950-55.
2 Dépôt des Engagés Volontaires Etrangères; worn 1940s.
3-4 1er Régiment Etranger; (3) has protective plate of thick plastic, both worn 1955-83.
5 1er Régiment Etranger d'Infanterie (REI); worn 1951-55.
6 Compagnie Montée d'Algérie du 1er REI; * worn 1934-38.
7 4è Bataillon, 1er REI; 1937-39.
8 Compagnie de Sapeurs Pionniers du 1er REI; 1937-40.
9 Compagnie Auto du 1er REI; 1933-40.
10 1er Bataillon du 3è REI; 1936-37.
11 Détachement Légion Etrangère Comores; worn 1973-76.
12 Détachement de la Légion Etrangère de Mayotte; 1976-83.
13 Compagnie de Pionniers de la Légion Etrangère; Mar-Dec 1968.
14 8è Compagnie du 3è REI; c1950.
15 Compagnie Disciplinaire des Régiments Etrangers; 1955-62.
16 Compagnie Disciplinaire en Extrême-Orient; 1948-54.
17 2è REI; 1957-83, second type.
18 4è Compagnie Mixte Portée du 2è REI; * worn 1941-43.
19 2è REI; 1946-56 – see also (17).
20 Régiment de Marche de Légion Etrangère en Extrême-Orient; worn 1945-46.
21 5è Bataillon du 3è REI; 1951-54.
22 6è Compagnie, 2è Bataillon du 3è REI; * worn 1946.
23 3è REI; 1928-43, first distinctive regimental insignia of Legion.
24 3è REI; 1945-46, fourth type.
25 7è Compagnie, 2è Bataillon du 3è REI; small type, see (29).
26 3è REI; 1945-46, second type.
27 7è Compagnie, 2è REI; 1965-67.
28 3è Bataillon de Marche de Légion Etrangère; 1962-64.
29 7è Compagnie, 2è Bataillon du 3è REI; c1945, large type – see (25).
30 Régiment de Marche de Légion Etrangère; 1943-45.
31 3è REI; 1945, second type.
32 3è REI; 1954-83.
33 Bataillon de Marche, 4è Demi-Brigade de Légion Etrangère; * worn 1947-49.
34 1ère Compagnie du Régiment d'Instruction de Légion Etrangère; worn in 1980s.
35 3è Compagnie du 4è REI; 1930s.
36 2è Bataillon du 4è REI; 1936-40.
37-38 4è REI; (37) * variant of (38), the insignia worn 1937-64.
39 5è Bataillon du 4è REI; worn in Indochina, May-November 1949.
40 2è Bataillon du 4è REI; 1947-51.
41-47 Company insignia of 4è REI: (41) 2è Compagnie Portée (CP), 1955-64; (42) 4è CP, 1956-64; (43) 6è CP, 1957-64; (44) 3è CP, 1955-64; (45) Compagnie régimentaire, 1958-64; (46) 1è CP, 1953; (47) Compagnie Mixte.
48 Legion paratroop unit; Djibouti.
49 5è REI; 1949-55, second type.
50 5è REI; 1955-63.

51 5è REI; * worn 1940-45, first type.
52 Train blindé du 2è REI en Extrême-Orient; * worn 1948-50.
53 3è Bataillon du 5è REI; 1945.
54 5è REI; variant of (49), 1949-55.
55-56 5è Régiment Mixte du Pacifique; 1963-83.
57 5è Compagnie du 3è REI; 1956.
58 1er Bataillon du 5è REI; * worn 1943-45.
59 1er Régiment Etranger de Cavalerie (REC); 1936-83.
60-64 Squadron insignia of 1er REC: (60) 2è Escadron, 1964; (61) 1er Escadron, 1952; (62) 3è Escadron, 1964; (63) 4è Escadron, 1970 version of insignia introduced 1963; (64) 5è Escadron, c1952; all insignia current, 1983.
65 16è Compagnie d'Entretien du Génie; 1946-52, first type.
66 21è Compagnie du 73è Bataillon
67 2è Compagnie de Réparation d'Engins Blindés de Légion Etrangère; 1951-55.
68 3è Compagnie de Réparation de Légion Etrangère; 1949-51.
69 Compagnie de Passage de la Légion Etrangère; 1948-55.
70 16è Compagnie d'Entretien du Génie; 1952, second type.
71 Bataillon de Légion Etrangère à Madagascar; 1957-62.
72 76è Bataillon de Génie Légion; 1951-55.
73 Groupement Frontalier Est Tonkin.
74 2è Compagnie Légionnaire de Transport du 519è Groupe de Transport; 1949-53.
75 68è Bataillon Vietnamien issu du IV/13 Demi Brigade de Légion Etrangère; 1953.
76 2è Compagnie de Transport et de Quartier Général; * worn 1951-54.
77 4è Compagnie Moyenne de Réparation de Légion Etrangère; 1951-55.
78 61è Bataillon Mixte de Génie Légion; 1971-83.
79 2è Compagnie de Réparation de Légion Etrangère; 1949-54.
80 21e Compagnie du 71e Bataillon du Génie; 1949-53, first type.
81 1er Bataillon d'Ouvriers du Service du Matériel; 1948-51.
82 21è Compagnie du 71è Bataillon du Génie; 1952-not known.
83 1er Bataillon de Réparation du Matériel; 1951-54.
84 3è Compagnie de Transport de Légion Etrangère du 516è Groupe de Transport; * worn 1950-53.
85 2è Compagnie de Transport de Légion Etrangère du 519è Groupe de Transport; 1950-not known.
86 34è Bataillon de Génie.
87 723è Compagnie de Réparation Auto; * worn 1946-52, second type.
88 1ère Compagnie de Réparation de Légion Etrangère; 1949-51.
89 40è Compagnie de Camions Bennes des Troupes Françaises en Extrême-Orient; 1951-55, second type.
90 21è Compagnie du 61è Bataillon du Génie; c1949-53, first type.
91 5è Compagnie Moyenne de Réparation de Légion Etrangère; 1951-55.
92 2è Compagnie du 74è Bataillon du Génie; 1951-52.
93 38è Compagnie de Camions Bennes; 1947-65.
94 1er Régiment de Marche de Volontaires Etrangers; * worn 1939-40.
95 2è Régiment de Marche de Volontaires Etrangers; 1939-40.

1-3 13è Demi Brigade de Légion Etrangère; three variations of the third type insignia, originated 1946. (2) is current.
4-5 Groupement Porté de la Légion Etrangère du Maroc; 1946-56.
6 21è Compagnie Portée (CP); worn 1954-1955.
7 22è CP; 1955-56.
8 23è CP; 1955-56.
9 6è Régiment Etranger d'Infanterie; 1941-55, second type.
10 2è Compagnie du Régiment d'Instruction de Légion Etrangère; 1980-83.
11 1er Escadron Saharien Porté de Légion Etrangère; 1961-63.
12 4è Compagnie Saharienne Portée Légion; 1956-63.
13-14 Promotion Narvik, Ecole Militaire Interarmes; insignia of a training establishment.
15 1er Régiment Etranger de Parachutistes (REP); 1951-61.
16 1er Bataillon Etranger de Parachutistes; * worn 1948-50.
17 1er Compagnie du 2è REP.
18 2è REP; 1963-83.
19 2è Régiment Etranger de Cavalerie; 1939-62.
20 Compagnie Saharienne Portée Légion; 1940-46.
21 3è Bataillon Etranger de Parachutistes; 1949-55.
22-23 Training establishment insignia: (22) Promotion Lieutenant Huard de Verneuil; (23) Promotion Général Rollet.
24 Cloth sleeve badge, 11th (Parachute) Division, which includes 2è REP.
25 Cloth sleeve badge of Legion units stationed in Sahara area.
26-27 Training establishment insignia: (26) Promotion Camerone; (27) Promotion Capitaine Danjou.
28-34 Beret badges: (28) Legion Cavalry, c1960-83; (29) 1er Régiment Etranger de Cavalerie (REC); (30) 2è Régiment Etranger d'Infanterie (REI), late 1950s-1960s; (31) 1er REI, 1958-60; (32) 2è REC, c1958-1960s; (33) 3è REI; (34) 4è REI, 1957-60.
35 Legion Medical Orderly; worn on neck of fob above unit insignia.
36 6è REI: miniature "former unit" badge, worn on shoulder board beneath grenade symbol.
37-41 British-made qualification badges for Polish Air Force in Britain, World War II. Basic designs date from 1919. Secured to left breast by nut and screw; chain hooks into left lapel. (37) Air Gunner, 1944 on; (38) Flight Engineer, 1944 on; (39) Air Gunner/Radio Operator, 1944 on; (40) Observer, 1919-42; (41) Air Gunner/Radio Operator, 1942-44.
42-87 US Army Collar Discs, 1907-36. Dress Regulations 1907 prescribed wearing of 1in (25mm) diameter bronzed discs on collar of khaki service tunic by all enlisted ranks. National or State discs worn on right; Arm of Service or Department on left. Matt gilded discs (76-87) were introduced in 1924. (42) National Army (draftees); (43) Signal Corps; (44) New York National Guard (NG); (45) District of Columbia NG; (46) Headquarters Unit; (47) Trains; (48) Regular US Army; (49) Florida NG in Federal Service, Regt No 1; (50) Georgia NG; (51) Kentucky NG; (52) National Army, see also

(42); (53) North Carolina NG; (54) Quartermaster Service; (55) not identified; (56) Infantry, Company H; (57) Artillery Band; (58) Medical Service; (59) US Reserve; (60) 3rd Infantry Regt, Company K; (61) 311th Regt; (62) 1st Infantry Regt, Company E; (63) Infantry; (64) Civilian Conservation Corps, worn on military-style uniform; (65) Infantry, Headquarters Company; (66) Regular Army, 346th Infantry Regt, Company B; (67) Infantry, Machine Gun Company B; (68) Air Corps; (69) Engineer Corps; (70) Regular Army, 59th Regt; (71/72) Motor Transport Officer, right and left collar badges respectively; (73/74) Motor Transport Enlisted Ranks; (75) not identified; (76) Regular Army, 7th Regt; (77) Artillery, Battery F; (78) Illinois NG; (79) Armored Corps, post-1936 variant; (80) Pennsylvania Military College; (81) Regular Army; (82) not identified; (83) Reserve Officer Training Corps; (84) Cavalry, Troop F; (85) Wisconsin NG in Federal Service, Regt No 120; (86) Coastal Artillery, Battery B; (87) US Army Band.

88 Enamelled badge with insignia of 93rd Infantry Division, US Army. The main strength of the Division in World War I was made up by the all-Negro 369th, 370th, 371st and 372nd Infantry Regiments. It was largely armed and equipped from French stock and, in honour of its distinguished service, has the French steel helmet as its insignia.

89 Rank chevrons, Corporal, US Army Air Corps.

90-104 Distinctive Insignia of the US Army Air Forces, World War II. These metal badges, resembling the shoulder sleeve patches shown on *pages 124-125,* were for wear on the shoulder straps. (90) 1st Air Force; (91) 2nd Air Force; (92) 3rd Air Force—note that this should show the "3" inclined to the right; (93) 4th Air Force; (94) 6th Air Force; (95) 7th Air Force; (96) 8th Air Force; (97) 9th Air Force; (98) 10th Air Force; (99) 11th Air Force; (100) 12th Air Force; (101) US Army Air Forces in Europe; (102) US Army Air Force; (103) 15th Air Force; (104) US Army Air Forces, second pattern, post-1942.

Metal badges, US Navy and US Marine Corps; including (top) insignia of the famous Treasure Island base in San Francisco Bay.

1 Qualification Badge, Pilot-Observer, Luftwaffe (Air Force).
2 West Wall Medal (*Schutzwallehrenzeichen*); for those concerned in construction of Germany's western defence fortifications.
3 Tank Battle Badge (*Panzerkampfabzeichen*) in bronze; to crews of support AFVs and Panzer-Grenadiers, for 50 tank battles. (See also 4, 6, 8, 17.)
4 Tank Battle Badge in bronze with gilt wreath; as (3), for participation in 100 tank battles.
5 General Assault Badge (*Allgemeines Sturmabzeichen*), black metal with silver wreath; for 25 assault actions (see also 11); worn on left breast pocket by men not entitled to Infantry Assault Badge (see 19) or the Tank Battle Badge (see 3, 4, 6, 8 and 17).
6 Tank Battle Badge in black metal with silver wreath; for tank crews (see 3, 4, for support troops) for 50 tank battles. (See also 8.)
7 Motor Torpedo Boat War Badge (*Schnellboot-Kriegsabzeichen*) in grey metal with gilt wreath, post-1943 issue; for crews or individuals, Navy, for distinction in combat.
8 Tank Battle Badge in silver and gilt; for tank crews, for 100 tank battles.
9 Special Badge for Single-Handed Destruction of a Tank (*Sonderabzeichen für das Niederkämpfen von Panzerkampfwagen durch Einzelkämpfer*); second class on silver corded cloth (first class—given after five second class—on gold cloth); worn right upper sleeve; for single-handed destruction of a tank without the use of an anti-tank weapon.
10 High Seas Fleet War Badge (*Flotten-Kriegsabzeichen*); for battleships' and cruisers' crews for distinguished single action or 13 weeks' active sea service; or to individuals wounded in action or surviving sinking.
11 General Assault Badge, white metal; first type, June 1940; for assault engineers (*Sturmpionier*) making three separate assaults.
12 Minesweepers, Anti-Submarine and Escort Vessels War Badge (*Kriegsabzeichen für Minensuch-, Ubootsjagd- und Sicherungsverbände*); for three sorties or one distinguished action.
13-14 Metal (13) and embroidered (14) Qualification Badge, Wireless Operator/Air Gunner, Luftwaffe.
15-16 Embroidered (15) and metal (16) Qualification Badge, Parachutist, Luftwaffe.
17 Tank Battle Badge in bronze; basic award for support troops.
18 Anti-Aircraft Artillery War Badge (*Kampfabzeichen der Flakartillerie*).
19 Infantry Assault Badge (*Infanterie-Sturmabzeichen*) in silver; also in bronze for motorised infantry.
20 Naval Combat Clasp (*Marine-Frontspange*), bronze; given after five awards of a naval war badge.
21-22 Qualification Clasp, Bomber Aircrew, Luftwaffe (*Frontflugspange*); awarded for specific number of operational sorties in bronze, silver (21) or gold (22).
23-24 German Cross (*Deutsches Kreis*) in silver (23) and in gold with brilliants (24); awarded in gold for bravery in action, in silver for distinguished leadership.

25-26 War Merit Cross (*Kriegsverdienstkreuz*), 2nd Class with Swords, with original paper container. (See 28, 30, 31.)
27 Sudetenland Medal (*Medaille zur Erinnerung and den 1 Oktober 1938*); for participation in the occupation of Czechoslovakian border regions, 1938.

*The awards and decorations won by Rear-Admiral (*Konteradmiral*) Hans Walter (1883-1950) in a career spanning two World Wars. Of special interest here is the Submarine War Badge of the Imperial German Navy (top centre, above Iron Cross), awarded in August 1918, when Walter commanded Submarine Flotilla* Flandern 1. *Full details of Walter's medals are given on* pages 69-70.

28 War Merit Cross, 1st Class with Swords, in original case.
29 Arm Shield commemorating Army's Crimea campaign of 1941-42. Eight shields to mark battles or campaigns, to be worn on the left upper sleeve, were authorised for issue between 1940 (Narvik) and 1945 (Lapland).
30-31 War Merit Cross, 2nd Class (30) and 1st Class (31); awarded with Swords – see (25-26) – only for active service.
32 Spanish Cross (*Spanienkreuz*) in bronze; for service in Spanish Civil War, 1936-39.
33-35 Wound Badges; in black (34) for one-two wounds; silver (35) for three-four wounds; gold for five-plus wounds; and (33) to mark attempt on Hitler's life in the "July Plot" of 1944.

36 Cross of Honour for Survivors and Bereaved of German Participation in Spanish Civil War (*Ehrenkreuz für Hinterbliebene deutscher Spanienkämpfer*).
37-39 Roll of Honour Clasps; "bars" to the Iron Cross for (37) Army; (38) holder of Iron Cross, 2nd Class, awarded World War I; (39) Luftwaffe.
40 Iron Cross, 1st Class; 1939 pattern.
41 Submarine Combat Clasp (*Uboots-Frontspange*) in silver; for multiple awards of Submarine War Badge.
42 Close Combat Clasp, Army (*Nahkampfspange des Heeres*) in silver; for 30 days' unsupported action.
43 Close Combat Clasp, Luftwaffe; silver centre, rest bronze; for 15 days' close ground combat.
44 Submarine Combat Clasp in bronze; see also (41).

Medals

J.D. Sheen

Military medals are the official marks of recognition and reward for bravery, for participation in a battle, campaign or war, or for outstanding services rendered by an individual. They are the outward sign to others that the recipient, the wearer, has merited distinction. Medals have been awarded since the development of organised armies, from Roman times and even before, but it was not until much later that they generally took the form of "badges" to be worn on the uniform, and later still (around the mid-19th Century) when, except in full dress, only the coloured ribbon was worn to mark the honour.

For reasons that will be made clear, British medals are among the most popular with collectors all over the world, and a great part of this summary will thus focus on British awards. However, Britain came fairly late to the custom of awarding medals: on his way into exile in July 1815, Napoleon I remarked that none of his British guards had medals, in contrast to the men of the French Army of the time, thus implying that they were mere recruits. Their commander had to assure the deposed Emperor that his escort was composed of veterans.

Although awards had been made by Queen Elizabeth I, for service against the Spanish Armada in 1588, and by Charles I for gallantry in the battles of the Civil War of the 17th Century, these were intended for commanders and their officers rather than for the rank and file. The veterans of the Peninsular and Napoleonic Wars of 1793-1815, like those mentioned above, received no recognition by the issue of medals until 1847, when survivors were retrospectively issued with the comprehensively named Military General Service Medal, 1793-1814.

WATERLOO MEDAL

There was an exception: the survivors of the Battle of Waterloo, 1815, were awarded a silver medal which was of the same design for all ranks. Issued in 1816, it bore on the obverse (the "front") the head of the Prince Regent and on the reverse the personification of "Victory". This was the British Army's first universal issue campaign medal and, an important point, the first to be "named"; ie, to have the name and unit of the recipient machine-stamped around the edge.

Today, some countries award medals in profusion; the United States of America is often cited as an example of over-generosity with awards. From the collector's point of view, of course, the fact that a medal is awarded in great numbers may make it less desirable because less rare (it will also be less expensive). But the value of the medal to its recipient is not necessarily less; it is simply that some countries are more ready than others to reward achievements that may not involve active service.

Some collectors acquire medals for their investment value and, of course, the finer and rarer the medal, the better the investment. But the majority of collectors do not cherish medals for their monetary value or their rarity (and the two things are generally synonymous) alone; many will be found to treasure a medal or group of medals (an array) with little value on the open market, perhaps because of what they know of the personal history of the recipient, perhaps because they were given for actions or campaigns which are of particular interest to the individual collector.

THEMES AND RESEARCH

This brings us to the important question of specialisation. Like other

Above: *As shown in this impressive array, the Victoria Cross takes precedence over all other British awards.*

Far left: *The German "manned-torpedo" pilot's exploits were rewarded (Inset) with the Knight's Cross of the Iron Cross.*

Left: *A Soviet-US ground inspection team in Germany, post-1945; the Russian Colonel wears the Order of the Red Star.*

Right: *Campaign medals mark out two NCOs of the Indian Army's 15th Lancers (Cureton's Multanis) as veterans, c1909.*

forms of collecting, medal collecting is more rewarding if undertaken on a thematic basis, rather than haphazardly. Because most British medals are named, it is possible to specialise in medals to men of particular regiments. One may specialise in gallantry medals, from the Victoria Cross (only for the very wealthy) downwards; in the medals awarded for one particular war; in campaign medals generally; or in the medals awarded during the reign of one sovereign. Then there are the medals awarded for long service, for good conduct, or to mark a Royal coronation or jubilee.

Medals are history. Each medal represents an achievement by one man, the recipient, and a named medal allows the joy of possession to be immeasurably increased by the fascination of research into the background of the recipient. Look, for example, at the arrays of medals at (1-8) on *pages 72-73* or at (1-6) on *pages 74-75;* here, study of Army Lists, regimental histories and other records that are available for consultation in the Public Record Office, London, in regimental museums and in public or specialist libraries, has allowed the military career of an individual to be traced from beginning to end. The value of the most common medal is thus increased for the collector himself as research brings to life the story behind his medals.

However, not *all* British medals are named. Some gallantry medals are not named—and the recipients of campaign medals for World War II generally felt that it was downright meanness on the part of the Government, rather than a prudent economy measure, that their medals were not named. The collector will be inclined to share this opinion. As a result, British campaign medals of World War II have no great value; an exception is the Air Crew Europe Star (see (13), *pages 74-75*).

Most other countries do not name their medals. The medals are issued with a citation or certificate which, however impressively printed, tends to be destroyed or lost in time, so that the recipient can no longer be identified. Identification is usually only possible in the case of a famous or particularly distinguished individual whose medals have been preserved as a group.

ADMIRAL WALTER'S MEDALS

An example of the way in which the career of an distinguished serviceman is expressed in his decorations is seen in the splendid array shown in the *Inset, pages 66-67.* Rear-Admiral (*Konteradmiral*) Hans Walter (1883-1950) joined the Imperial Navy as a cadet in 1902 and in August-December of that year participated in the blockade of Venezuela aboard the corvette *Stosch.* He was promoted Lieutenant (*Leutnant*) in 1905, and in 1907, as a staff officer on the battleship *Preussen,* he was awarded the Prussian Order of the Royal Crown, Fourth Class (fifth from left in array). Promoted First Lieutenant (*Oberleutnant*) in 1908, he served on the light cruiser *Geier* and heavy cruiser *Gneisenau* before becoming a Submarine School instructor in 1913, when he was awarded the Colonial Medal with the bar "Venecuela" (eighth left). He became a Lieutenant-Commander (*Kapitänleutnant*) in 1913, while commanding the submarine *U.3.*

During World War I, Walter commanded the submarines *U.17* and *U.52,* winning the Knight's Cross with Swords of the Royal House Order of the Hohenzollerns (second left) for sinking the British light cruiser *Nottingham* and, in November 1916, the *Pour le Mérite,* Germany's highest award for valour, for sinking the French battleship *Suffren.* From October 1917 he com-

Far left: *This German air "ace" of World War I wears the uniform of Turkey, then Germany's ally, but at his throat, its black and silver ribbon hidden by his tunic's collar, is Germany's highest award for gallantry in action,* Pour le Mérite, *popularly known as the "Blue Max". Among his other decorations is the Iron Cross, First Class, worn always pinned to the tunic, without a ribbon.*

Left: *Unlike Britain and some other major military nations, the United States awards medals for proficiency. Designed by the US Mint and authorised in 1932, the Expert Pistol Shot Medal of the US Navy is awarded to all USN and USNR personnel who attain a specified standard of marksmanship.*

Right: *The collection of ribbons is an acceptable alternative to medal collecting for those of limited means. As the arrays of General L.I. Lemnitzer, USA, (left), and Admiral H.D. Felt, USN, show, there is certainly no lack of American material in this field.*

manded a submarine flotilla; his other wartime decorations were the Iron Cross, First and Second Class, the Frederick Augustus Cross, First Class, of the Grand Duchy of Oldenburg (third left), and the Austrian Military Cross of Merit, Third Class with War Decoration (extreme right). He was awarded the Submarine War Badge of the Imperial Navy in August 1918.

Walter ended World War I as one of Germany's most successful submarine commanders, having sunk a battleship, a light cruiser, a submarine and 30 merchant ships totalling some 100,000 tons. He was promoted Commander (*Korvettenkapitän*) in 1921 and retired in 1929 as a Captain (*Kapitän zur See*). After receiving the Active War Service Cross (fourth left) for his World War I service, in 1935, he returned to serve in the Navy of the Third Reich (*Kriegsmarine*) in 1936. Promoted Rear-Admiral (*Konteradmiral*) in 1941, he received the Forces Long Service Award in Gold and a similar meritorious service award (sixth and seventh left) and retired finally in 1942.

Although such a distinguished array will be beyond the reach of most collectors, it is possible, by diligent research, to build up an equally detailed and fascinating history—especially with named British medals—from an array of much more common medals; even from the most common British group, the 1914-1915 Star, War Medal and Victory Medal (see *pages 72-73*), which are usually known as "Pip, Squeak and Wilfrid" from a popular comic strip of the time.

EUROPEAN AND US MEDALS

Probably the best-known of European medals is the Iron Cross of Prussia or, later, Germany. Introduced in 1813 by King Frederick William III of Prussia, it took three basic forms: the Grand Cross (rarely given) and the First and Second Class, the former always worn without a ribbon. It is found in versions dated 1813, 1870, 1914 and 1939. The first three are all of the same basic shape, but Hitler's re-institution of the award in 1939 introduced several variations: the rather larger Knight's Cross and the Knight's Crosses with Oakleaves; Oakleaves and Swords; Oakleaves, Swords and Diamonds; Golden Oakleaves with Diamonds (awarded only once); Grand Cross (awarded only once). Examples of Iron Crosses of various types are shown on *pages 66-67* and *76-77,* while more will be seen with other German medals and the awards of German states on *pages 80-81* and *82-83.*

As has been stated, France was more generous than Britain in the award of medals from the 19th Century onward. France's premier award, the Legion of Honour, was instituted in 1802; since World War I it has, when awarded for service in the field, carried with it the mandatory award of the *Croix de Guerre* (see (16), *pages 76-77,* where several other French medals are shown), established in 1915. The *Médaille Militaire* was established in 1852 and could be won by Other Ranks as well as Officers. France awarded significantly more campaign medals than Britain during World War I, adding such special medals as those for public service in wartime (*Médaille de la Reconnaissance Française*), for victims of the invasion (*Médaille des Victimes des de L'Invasion*), for prisoners-of-war who escaped (*Medaille des Evadés*), for volunteering for service and for wounds or illness sustained through combat. This was in marked contrast to Britain: gallantry awards apart, British servicemen could gain no more than three medals for service in 1914-1918.

The victorious Allied nations of World War I issued a Victory Medal with a standard ribbon but with designs varying from country to country. A number of these are shown on *pages 76-77:* particularly notable is the Victory Medal issued by the United States of America (shown at (42) on *pages 76-77*), for which no less than 38 bars, 19 for the US Army and 19 for the US Navy, were authorised. A wide selection of US medals is shown on *pages 78-79,* spanning the period from World War I to the Vietnam War. Since most were awarded in great numbers, US campaign medals form an attractive and comparatively inexpensive field of specialisation for the novice collector.

BUYING AND CLEANING

Broadly speaking, the value of a medal will depend on its rarity and on its condition. A very rare medal, of course, will be acceptable to the collector in almost any condition, but for the general run of, say, British campaign medals, the collector will not wish to have examples that fall below the condition described by reputable dealers as "VF" (very fine; ie, with some signs of wear but not badly worn).

In the case of British campaign medals before World War II, which can be identified to an individual and a unit, value will be influenced by these factors. If research shows that the individual to whom the medal was awarded was, for example, killed on the first day of the Somme offensive in 1916 or in some other notable action, the value will be increased. Generally, medals awarded to Officers are valued higher than those awarded to Other Ranks and medals awarded to men of elite units, Cavalry and Guards, higher than those to men of Infantry Regiments or Corps. In the case of gallantry medals, which may not be

named, the value will be influenced by the provenance, the presence of a citation. In almost all cases, a group of medals awarded to the same individual will be of greater value than that of the medals separately: it is a good rule never to break up a group.

The variations in value depending upon the naming of the medal lead to consideration of two dangers facing the collector: the renamed medal and the copy. Renaming may have been carried out with no intent to defraud: a soldier who had lost a campaign medal might buy another, possibly from a pawn shop, and have the original name, number and unit erased and replaced by his own. There are several ways to tell if a medal has been renamed: the style of lettering may be unorthodox or the dimensions of the medal may have been fractionally reduced; if the work has been crudely done, then the medal may no longer form a true circle. What is almost impossible is to tell whether the renaming is "genuine", as described above, or "false", when the name has been changed for one of higher rank in a more prestigious unit.

BEWARE OF COPIES

The collector must also beware of reproductions; in recent years it has become profitable for unscrupulous persons to produce copies not of the major gallantry awards—which tend to be so costly that no one would buy without the most satisfactory provenance—but of the scarcer campaign medals, notably the British Air Crew Europe Star, mentioned above, and various combat awards and decorations of the Third Reich, which are in increasing demand among collectors. Only a thorough knowledge of the subject will help the collector: study the genuine medals in reference works and museums and be alert for any variation in design. In the case of the Air Crew Europe Star, for example, the tops of the "V" and "I" in the Royal Cypher touch in the genuine medal but have a space between them on the copy. The best advice is to be wary of remarkable bargains and to buy from a reputable dealer who will vouch for the authenticity of his stock and will offer a money-back guarantee if any irregularities should be found.

Since condition is so important, should medals be cleaned? Older medals may already be well worn, since abrasive metal polish was commonly used in the earlier part of this century, and the further use of abrasive cleaners should certainly be avoided. If you must clean, it is wise to seek the advice of an expert collector, but a few hints may be given here. Remember that medals are made of various kinds of metal: silver, cupro-nickel, bronze and other alloys. For silver medals, jewellers' rouge may be used, or perhaps a very light cleaning with ammonia or even a "silver dip" solution. For other metals, a wash in soapy water is probably best. The best advice is to keep cleaning to a minimum.

Again, some collectors wish to retain the "original" ribbon, however worn—overlooking the fact that the original owner may have changed the ribbon a number of times. In most cases, a length of new ribbon can be obtained cheaply from a dealer (check on the availability, of course, before discarding the old ribbon), and the substitution of a new ribbon will improve the display without lessening the value of the medal. If the medal has enamelled parts, as many continental orders do, these can be lightly cleaned with ammonia. They must be carefully dried, which is best done with a hair-dryer.

COLLECTING RIBBONS

Since ribbons are generally easily and cheaply available, it is possible to build up an interesting collection of ribbons alone. An attractive display may be made by mounting a large map of the world bordered by campaign ribbons, with a thread from each leading to a coloured pin on the map marking where the medal was won. Small captions with dates and other information may be added above the ribbons. Another less costly alternative to the collection of medals proper is that of collecting miniatures, exact copies made for wear on mess dress. These may be purchased from the makers or from military tailors, and although they have no extrinsic value they make an attractive display. (They may, of course, also be used to supplement a display of medals proper.)

So far as display is concerned, the most satisfactory method for a large collection is to use purpose-built medal cabinets with shallow, lined drawers; heavy and lockable, these provide some security as well as dust-free storage and easy access for the collector. They are expensive however, and shallow trays lined with soft material and provided with a top cover of some kind are a cheaper alternative. Display in glazed picture-frames, with the ribbons pinned to a suitable backing material, will make an impressive addition to the interior decoration of the collector's home: the disadvantage is that it is likely to be a lengthy process to add medals to the collection or to substitute better specimens for worn ones. The simplest method is to purchase a loose-leaf album of pocketed plastic sheets, of the kind used by many dealers to display their stock at militaria fairs.

1-8 These medals would be highly prized by any collector: all were awarded to a long-serving Bandsman in the Devon Regiment. Pride of place is taken by the Distinguished Conduct Medal (1), the first award to the Devon Regiment in World War I and one of the first to the entire British Expeditionary Force of 1914. It was won for bringing out wounded under fire (it has long been the practice for British Army Bandsmen to act as stretcher-bearers). The remaining medals in this scarce and valuable group are: (2) The India Medal 1895, with bar Punjab Frontier 1897-98; (3) Queen's South Africa Medal 1899 (reverse shown), with bars Belfast, Relief of Ladysmith, Elandslaagte; King's South Africa Medal 1901-1902 (the bar for 1901 was given to all men with 18 months' service pre-dating 1 June 1902; the bar for 1902 to all serving in South Africa after 1 January 1902; only nurses received the medal without a bar); (5) 1914 Star ("Mons Star") and bar (see inset photograph, right, for inscribed reverse of this medal); (6) British War Medal 1914-1920; (7) Victory Medal 1914-1918; (8) Army Long Service and Good Conduct Medal.

9 Tibet Medal with bar Gyantse; awarded for service between 13 December 1903 and 23 September 1904. Tibet Medals to Europeans are rare: this one was awarded to an Artillery sergeant attached to an Indian Army unit.

10 Africa General Service Medal 1902-1956 with bar for service in Somaliland 1908-1910.

11 Third China War 1900 with bar Relief of Pekin; awarded to a Royal Navy stoker for service in a supply ship during the "Boxer Rebellion", from 10 June to 3 December 1900.

12 Army Long Service and Good Conduct Medal; Victorian issue, compare with 20th-century issue shown at (8).

13 Military Cross; widely issued in World War I, the MC was awarded both for gallantry and in the King's Birthday and New Year Honours Lists. Those for specific acts of gallantry are more valued than general issues.

14 India General Service Medal 1908-1935 with bar Afghanistan N.W.F. (North West Frontier) 1919.

15 Reverse of Victory Medal 1914-1918; note the bilingual inscription: for South African recipients.

16 Colonial Auxiliary Forces Long Service Medal 1899-1921; this was issued to a private in the Falklands Islands Defence Force and, like the Falklands badge attached to the ribbon, is named and numbered to the recipient.

17 Territorial Force War Medal 1914-1919.

18 Mercantile Marine War Medal 1914-1918; reverse shown, obverse as (6).

19 Army Medal for Meritorious Service, on ribbon of post-1917 pattern.

20-24 A rare and most interesting World War I group, awarded for service aboard "The Terror of the Danube", a Royal Navy tug with improvised machine-gun and

torpedo-tube armament that took part in a hazardous action culminating in the sinking of an Austrian warship on the River Danube. Notable are (20) Distinguished Service Medal, authorised in 1914 for bravery by petty officers and men of the Royal Navy and equivalent ranks of the Royal Marines; and (24) the Serbian Medal for Bravery (called the "Obilitch Medal", since it bears a portrait of the national hero) in silver. At (21) is the 1914-1915 Star, awarded for service in any theatre of war in 1915, but not awarded to holders of the 1914 Star (5), which it resembles in all except date scroll.

25 A rare item: medallion issued by the Government of Ceylon (now Sri Lanka) to all Ceylonese volunteers in World War I.

26 Canadian Memorial Cross, or "Mother's Cross", issued to the next-of-kin of Canadian servicemen who died in World War I; these are officially named and numbered.

27 General Service Medal 1918-1964 with bar Iraq (1919-20).

28 Military Medal and bar; this gallantry award (with bars for further acts) for NCOs and men of the British Army was instituted in March 1916.

29 Royal Air Force Long Service and Good Conduct Medal; this George V medal to a comparatively young service is quite rare.

30 Territorial Long Service and Good Conduct Medal, pre-World War I issue.

31 Military Long Service and Good Conduct Medal; see (8): the change of name and style, with Regular Army bar, came about in 1930.

32-36 A group awarded to a private in the Royal Army Medical Corps, who did not survive World War I. Notable is (32) Death Plaque, sent to next-of-kin with a letter of condolence from the King – although named, it does not bear number or unit and is thus difficult to research if found alone. Also (33) Military Medal; (34) "Mons Star" and bar; (35) British War Medal; (36) Victory Medal.

Reverse of 1914 Star ("Mons Star"), showing inscribed name of recipient. Inscription to a man of Collingwood Battalion, Royal Naval Division, makes this common medal a rarity.

(Left) South Atlantic Medal, for service (in the war zone, as the rosette shows) in the Falklands Campaign, 1982. (Right) Queen Elizabeth II Jubilee Medal, 1977.

1-6 This interesting group allows the researcher to trace the career of the recipient. He gained the 1914-1915 Star (1) as a Private in the Ceylon Planters Rifle Corps, but ended World War I, with the British War Medal 1914-1920 (2) and Victory Medal 1914-1918 (3), as a Captain-Pilot, Royal Flying Corps. He rejoined his old unit in the ranks, receiving the Colonial Auxiliary Forces Long Service Medal (6) and King George VI's Coronation Medal, 1937 (4). Finally, as a Major in the Ceylon Planters Rifle Corps, he received the Territorial Decoration with bar Ceylon (5).

7 Distinguished Conduct Medal, awarded to a Sergeant, Royal Army Medical Corps, for bringing in wounded under fire, 1914.

8 Special Constabulary Long Service Medal; the original issue, authorised in August 1919, with the bar The Great War 1914-18.

9 Army Long Service and Good Conduct Medal; George V issue, pre-1930 (see 31, *pages 72-73*).

10 Distinguished Flying Cross; gallantry award for officers and warrant officers of the Royal Air Force, authorised in 1918. This one was awarded to a Spitfire pilot of World War II.

11 1939-1945 Star, awarded for active service (time and conditions varying) in World War II. This and the other World War II campaign stars were issued unnamed.

12 Atlantic Star, awarded to members of all arms for service afloat in Home Waters, the Atlantic and Russian Convoys (varying qualifications), World War II.

13 Air Crew Europe Star, awarded for operations over Europe from UK bases, 3 September 1939 to 5 June 1944; the rarest of the World War II campaign stars.

14 Africa Star, awarded for service in North Africa (including Malta) or the Mediterranean (Royal or Merchant Navy), 10 June 1940 to 12 May 1942.

15 Italy Star, awarded for service in the Italian theatre, including Greece and Mediterranean islands, 11 June 1943 to 8 May 1945.

16 France and Germany Star, awarded for service in France, Holland, Belgium and Germany, between 6 June 1944 (ie, "D-Day") and 8 May 1945.

17 Burma Star, awarded for service in Burma, from 11 December 1941 to 2 September 1945.

18 Pacific Star, awarded for service in the Pacific theatre, 8 December 1941 to 2 September 1945.

19 India Service Medal, awarded for three years' non-operational service with Indian forces during World War II.

20 Southern Rhodesia Service Medal; a very rare medal, awarded only for World War II service in Southern Rhodesia.

21 South African Medal for War Service (reverse shown), awarded for at least two years' service in a voluntary, unpaid organisation (eg, Red Cross) from 6 September 1939 to 15 February 1946.

22 Distinguished Flying Medal, the equivalent of the DFC (10) for gallantry in combat by Air Force NCOs and men.

23-25 Trio of World War II medals awarded to a Canadian soldier. The Defence Medal (23) and War Medal (24) are exactly similar to the British issues, see (32) and (33), in appearance – but are struck in ·800 fine silver rather than cupro-nickel. The Canada Volunteer Service Medal (25) was authorised in 1943 for volunteers with at least 18 months' service; the maple leaf bar was worn only by those who served abroad.

26 Africa Service Medal, authorised in 1943 for all South African servicemen who enlisted before 13 May 1943. Unlike (21), it is named on the rim: for non-white recipients, the name has the prefix "C" for coloured or "N" for native.

27 Australia Service Medal, awarded to all Australian personnel, including civilians, who served for 18 months during World War II.

28 Naval Long Service and Good Conduct Medal, Elizabeth II issue; named to a Chief Petty Officer of HMS *Eagle*, one of four Royal Navy aircraft carriers in the Korean War.

29 Campaign Service Medal, awarded from 1962 onward, with the bars Radfan (1964) and South Arabia (1964-67).

30 Korea Medal, awarded to British Commonwealth Forces who served in South Korea, 2 July 1950 to 10 June 1953. This is the Canadian issue, in silver rather than cupro-nickel (note CANADA beneath queen's head): 27,500 of these were issued.

31 United Nations Service Medal, instituted in 1950 and issued to all servicemen of all nations who served in Korea. The medals, rather poorly made in bronze alloy, all have the bar Korea.

32 Defence Medal, awarded to both military and Civil Defence personnel for service (generally three years at home or six months overseas) in World War II.

33 War Medal, awarded to all full-time military personnel with at least 28 days' service during World War II.

34 General Service Medal, awarded 1919-1964 for campaigns which had no individual medal: this one has the bar Palestine 1945-48 and is the second George VI issue (the first issue shows the monarch without a crown). The inclusion of a named General Service Medal can increase both the interest and value of an otherwise anonymous World War II group by allowing it to be researched.

Medals of World War I: Austria, Belgium, France and Others

1 Medal for Bravery, Austria; silver, issued also in gold and bronze. The reverse bears the two-line inscription "DER TAPFERKEIT" ("bravery" or "gallantry").

2 Iron Cross of Merit, Austria; mainly issued to personnel of services medical, railway or post office corps. The ribbon for peacetime awards is plain red.

3 Wounded Troops Medal, Austria; issued from 1917. A red stripe was added to the ribbon for each wound up to a maximum of five.

4 Karl Truppenkrieg Medal, Austria; instituted in December 1916 and awarded for a minimum 12 months' field service and participation in at least one battle.

5 Commemorative War Medal, Austria; issued to all personnel who served in World War I. In 1933, crossed swords in gilt, as seen, were authorised for wear by all those who fought at the front, were wounded or became PoWs in honourable cirumstances.

6 Frederick August 1914 War Service Cross, Oldenburg; German State medal for meritorious service, issued with bar VOR DEM FEINDE ("In the face of the enemy").

7 Commemorative War Medal, Belgium; authorised in July 1919 for all who served in the Belgian Army in World War I.

8 Croix du Feu, Belgium; authorised as late as 1934 for award to all who served a specified time in a combat zone, World War I.

9 Yser Medal, Belgium; authorised in 1918 for all Belgian and Allied troops who took part in the action on the River Yser in October 1914. From 1934 onward, the medal was issued as a cross, with four arms added to the design.

10 Liége Medal, Belgium; issued to commemorate the defence of the city of Liége in 1914, this is the only non-official medal authorised to be worn on Belgian uniform.

11 Queen Elizabeth's Medal (*"de la Reine Elisabeth"*), Belgium; instituted in 1916 for award to women who aided Belgian troops or civilians in wartime. Medals with a red cross within the wreath—like the one shown here—were specifically for aid to sick or wounded servicemen.

12 Croix de Guerre, Belgium; instituted in October 1915, some six months after France's authorisation of (16). Awarded for bravery or, sometimes, long service at the front, and to those mentioned in despatches. The palm leaf on the ribbon indicates subsequent mention(s).

13 Victory Medal, Belgium; the ribbon of this medal was standard for all World War I allies, but there were variations in the design both on the obverse—see (14) and (42)—and reverse—see (39).

14 Victory Medal, France; authorised in 1922 for World War I service.

15 Medal for Medical Assistance, France; there are a number of French awards for brave deeds and saving lives, all with the blue/white/red ribbon.

16 Croix de Guerre, France; instituted in April 1915 for award to those mentioned in despatches. The bronze star on the ribbon denotes that the recipient of this medal was mentioned in despatches at

Brigade level or below. Other emblems are a bronze laurel branch (or "palm") for mention in an Army despatch; a silver-gilt star for Army Corps mention; a silver star for Divisional mention. More than one such emblem may be worn on the ribbon of a single medal.

17 Medal for Escaped Prisoners (Médaille des Évadés), France; instituted in 1926 for PoWs who escaped from enemy hands.

18 Combatants' Cross (Croix du Combattant), France; instituted in 1930 for all those who served in the French armed forces.

19 Medalet of a type often found attached to the ribbon of the Combatants' Cross (18).

20 Commemorative Medal of the Great War (Médaille Commemorative de la Grande Guerre), France; authorised in 1920 for all servicemen and some civilian auxiliaries of 1914-18.

21-22 Cross of Honour, Germany instituted in 1934 to mark World War I service. Issued in three categories: (21) non-combatants; (22) bronzed iron, with swords, for combatants; and in blackened iron for the next-of-kin of servicemen who fell in 1914-18.

23-24 Iron Cross, Second Class; Prussia; obverse (23) and reverse (24) – the latter with non-official pins, indicating other awards, on the ribbon.

25-28 Array of German medals for World War I, including Iron Cross, 2nd Class (25) and Cross of Honour (27), as well as two German State decorations.

29 Cross of Valour, Poland; for gallantry, instituted 1920.

30 Commemorative Medal of thanks to German servicemen issued by the city of Frankfurt.

31 Medal of St George, Imperial Russia; a bravery award for both soldiers and civilians.

32 Turkish Star ("Gallipoli Star"); awarded to Turkish soldiers and their allies for the Gallipoli campaign, World War I.

33 War Cross, Czechoslovakia; for military bravery, instituted 1918.

34 Iron Cross, First Class, Prussia; the reverse is plain.

35 War Commemorative Medal, Serbia; for all who served, 1914-18.

36 Commemorative Medal for Combatants, Hungary; instituted 1929, with a non-combatants' version without crossed swords.

37 United Italy Medal; issued for World War I service in 1922, but originally issued in 1883 to commemorate the 1848-70 unification of Italy.

38 Medal for Volunteers in 1915-18, Italy; instituted in 1923.

39 Victory Medal (reverse); Italy.

40 War Cross, Italy; instituted in 1918, for bravery or minimum of one year's service in combat zone (reverse shown).

41 War Medal, Italy; instituted in 1920 and supposedly made of bronze captured from the enemy.

42 Victory Medal, USA; the most interesting of the Victory Medals, since 38 area or battle bar awards were authorised.

43 Commemorative War Medal, USA; one of several semi-official issues, in this case presented by the State of Oregon for service in 1917-1918.

Medals of the USA: World War I to Vietnam

1 Exceptional Civilian Service Award, Department of Defense; one of several awards that may be made to civilian employees of the US services – see also (40).
2 United Nations Service Medal; award of this UN medal to all US personnel who served in Korea, 1950-54, was authorised 1951.
3 Joint Service Commendation Medal; authorised 1963 for servicemen distinguishing themselves while serving with a Joint Task Force, Command or Control Group.
4 Purple Heart; originally instituted 1782, revived 1932, this is now awarded only for death or wounds by enemy action.
5 Selective Service Medal, World War II; authorised 1945.
6 Humanitarian Service Medal; authorised 1979 for award to US servicemen participating in military operations of a humanitarian nature, such as famine relief.
7 Women's Army Corps Service Medal; authorised 1943 for women who served in the Women's Army Auxiliary Corps, 20 July 1942-31 August 1943.
8 National Defense Service Medal; authorised 1953, for any period of honourable service in US armed forces, June 1950-July 1954. Reinstituted for service beginning 1 January 1961.
9 European-African-Middle Eastern Campaign Medal; authorised 1946 for service in these theatres, 7 December 1941-8 November 1945.
10 American Defense Service Medal; authorised 1941 for minimum one year's service in armed forces, between 8 September 1939 and 7 December 1941.
11 World War II Victory Medal; authorised 1945 for at least one day's service, 7 December 1941-31 December 1946.
12 Asiatic-Pacific Campaign Medal; as (9), 7 December 1941-2 March 1946.
13 American Campaign Medal; authorised 1946 for service outside the United States or one year's aggregate service within, as (12).
14 Armed Forces Expeditionary Medal; authorised 1962 for at least 30 days' service in an operation for which no other service or campaign medal is authorised.
15 Medal for Humane Action; authorised July 1949 for those assigned to Berlin Airlift for at least 120 days, 26 June 1948-30 September 1949.
16 Army of Occupation Medal; for Army and Air Force personnel with at least 30 days' consecutive service in an Occupation Zone after World War II.
17 Vietnam Service Medal; authorised 1965 for at least 1 day's permanent or 30 days' temporary duty in Vietnam.
18 Korean Service Medal; authorised 1950 for at least 30 consecutive days' service in Korean Theater, 27 June 1950-27 July 1954.
19 Navy Achievement Medal; authorised 1967 for professional or leadership achievement by officers (below Lt-Cdr/Major) and men of US Navy and Marine Corps.
20 Navy and Marine Corps Medal; authorised 1942 for gallantry not involving actual conflict with an

enemy. Won by the future President John F. Kennedy in World War II.

21 Navy Cross; established 1919, and since 1942 the US Navy's second-highest gallantry award.

22 Army Good Conduct Medal; authorised 1941 for US Army enlisted men for 3-year periods of "honor, efficiency and fidelity". Succeeding awards denoted by bronze, silver and gold clasps.

23 Armed Forces Reserve Medal; authorised 1950 for reservists with 10 years' satisfactory service. Reverse has six different designs, according to arm; see (30, 35).

24 Antarctica Service Medal; authorised 1960 for servicemen and civilians serving in Antarctica or supporting operations there after 1 January 1946.

25 United States Army Reserve Medal; authorised 1960 for meritorious service in the US Army Reserve.

26 Distinguished Service Medal (Army); authorised 1918, the highest US award for services not involving personal gallantry. See also (33).

27 Air Reserve Forces Meritorious Service Award; authorised 1964 for NCOs with 4 years' exemplary service.

28 Good Conduct Medal (Air Force); authorised 1963, qualifications as (22). Succeeding awards: bronze and silver oak-leaf clusters.

29 Combat Readiness Medal; authorised 1964 for Air Force personnel with a 4-year period of service as combat-ready aircrew or missile-launch crew.

30 Armed Forces Reserve Medal; reverse of Air Force version.

31 Air Force Commendation Medal; authorised 1958 for servicemen distinguished by meritorious service in any capacity with the Air Force after 24 March 1958.

32 Air Force Cross; authorised 1960 for extraordinary heroism in action by any person of the Air Force.

33 Distinguished Service Medal (Air Force); authorised 1918, as (26).

34 Marine Corps Good Conduct Medal; authorised 1896 and currently awarded to enlisted men for 3-year periods of exemplary service.

35 Armed Forces Reserve Medal; reverse of US Marine Corps version, see (23).

36 Naval Reserve Meritorious Service Medal; authorised 1959 for four consecutive years' satisfactory service in Reserve.

37 Navy Good Conduct Medal; authorised 1869 for enlisted men with four years' continuous active duty, good military behaviour and leadership.

38 Navy Expeditionary Medal; authorised 1936 for naval personnel who have landed on foreign soil and engaged in operations against armed opposition for which no campaign medal is authorised.

39 China Service Medal; authorised 1942 for Navy and Marine Corps personnel serving in China, 7 July 1937-7 September 1939. Awarded again for service in China, Taiwan and Matsu Straits, 2 September 1945-1 April 1957.

40 Superior Civilian Service Medal; Department of the Navy; an award for civilian employees, as (1).

Medals of Imperial Germany and the Third Reich

1 Charles Frederick's Military Order of Merit (*Medaille des Militärischen Karl Friedrich Verdienstordens*) in silver; Grand Duchy of Baden.
2 Active War Service Cross (*Ehrenkreuz des Weltkrieges*), for service during World War I; Germany.
3 War Assistance Cross (*Kriegshilfekreuz*), for voluntary medical services, 1914-16; awarded by the Grand Duchy of Baden.
4 Hanseatic Cross (*Hanseatenkreuz*), awarded for service in World War I by the City of Lübeck, a free Hanseatic city within the German Empire until incorporated into Schleswig-Holstein in 1937. See also (19).
5 Frederick Francis-Alexandra Cross (*Friedrich Franz Alexandra-Kreuz*), World War I; Grand Duchy of Mecklenburg-Schwerin.
6 War Service Cross for Women and Girls (*Kriegsverdienstkreuz für Frauen und Jungfrauen*), with the inscription "For Devoted Service During the War" (World War I); Duchy of Brunswick.
7 War Service Cross (*Kriegsverdienstkreuz*); awarded by the Grand Duchy of Baden.
8 Military Aviator's Commemorative Badge (*Erinnerungsabzeichen für Militärflugzeugführer*), awarded by the German Air Service from 1914 onward.
9 War Decoration (*Kriegsehrenzeichen*) for public service, with the inscription "For Wartime Welfare Work"; Grand Duchy of Hesse.
10-13 In action in World War I, Second Lieutenant (*Leutnant*) Bernhard Müller carried his Iron Cross, Second Class (13), Visiting Card (11) and Identity Disc (12) in a Breast Pouch (10). A bullet struck the pouch and penetrated its contents with the effect seen here; unhappily, even this protection could not prevent Müller from receiving a mortal wound.
14-16 Iron Cross, First Class, World War I (16), with its original presentation box (15) and the cardboard case in which the box was packed (14). Note the inscription: "Only to be opened by the recipient".

Uniform tunic of a General, Army of the Third Reich. Beneath the right breast is the embroidered insignia of the German Cross (Deutsches Kreuz) *in Silver; on the left breast pocket, the Iron Cross, First Class (1914).*

17 Cross of the Order of Military Merit (*Militärverdienstkreuz*), Second Class with Swords; awarded by the Kingdom of Bavaria.

18 Frederick Cross for War Service (*Friedrichs-Kreuz*); Duchy of Anhalt.

19 Hanseatic Cross, as (4); Bremen, during World War I a "Republic and Free Hansa City" within the German Empire.

20 Anna Louisa Service Decoration (*Anna-Luisen-Verdienstzeichen*) for World War I; Principalities of Schwarzburg-Rudolstadt and Schwarzburg-Sondershausen, incorporated into the State of Thuringia in 1920.

21 Frederick and Matilda Medal (*Friedrich-Bathildis Medaille*); Principality of Waldeck.

22 War Service Cross (*Kriegsverdienstkreuz*) of 1914; Principality of Reuss, a state which was incorporated into Thuringia in 1920.

23 Cross of Honour of the Princely House Order of the Hohenzollerns (*Fürstlich Hohenzollernsches Ehrenkreuz*), First Class with Swords; Principality of Hohenzollern, Prussia.

24 William Cross (*Wilhelmskreuz*) with Crown, 1915; Kingdom of Württemberg.

25 Knight's Cross with Swords of the Royal House Order of the Hohenzollerns (*Hausorden von Hohenzollern*); Prussia; see also (23).

26 Array of decorations awarded (unless stated otherwise)by the Kingdom of Saxony. Left to right: Knight's Cross of the Military Order of St Henry (*Militär St Heinrich Orden*); Knight's Cross with Swords of the Order of Civil Merit (*Zivilverdienstorden*); Knight's Cross, First Class with Swords, Order of Albert the Valiant (*Albrechts-Orden*); Iron Cross, Second Class (1914), Prussia; as (23), Third Class with Swords; Meritorious Service Cross for Officers (*Dienstauszeichnungskreuz für Offiziere*).

27 Miniatures of medals (order changed) shown at (26); these were normally worn with mess dress or on civilian formal dress.

28 Commemorative Cross for Nursing (*Erinnerungskreuz für Krankenpflege*); Kingdom of Saxony.

29 Military Cross of Honour for an Heroic Action (*Kreigsehrenkreuz für heldenmütige Tat*), 1914-18; Principality of Lippe.

30 War Service Cross (*Kriegsverdienstkreuz*), Second Class; Duchy of Brunswick.

31 War Service Cross (*Kriegsverdienstkreuz*); Principality of Lippe.

32 Devoted Service Cross (*Kreuz für treue Dienste*); Principality of Schaumburg-Lippe.

33 Knight's Cross of the Order of the Zähringen Lion (*Orden vom Zähringer Löwen*), Second Class with Swords; awarded by the Grand Duchy of Baden.

34 Knight's Cross of the Order of Albert the Valiant (*Albrechts-Orden*), Second Class with Swords; awarded by the Kingdom of Saxony. See also (26).

35-38 German medal ribbons of the World War I period. Note enamel wreaths and gilt trophies marking awards with War Decoration or Swords of various classes.

1 An array of medals awarded to an Officer of the Imperial German Army for service in both war and peace. Prominent among them are: (left) Iron Cross, Second Class, 1914; (second left) the fairly rare Bavarian Order of Military Merit, Fourth Class with Crown and Swords; (third left) War Service Cross, 1914, Mecklenburg-Schwerin; (seventh left) Cross of the Order of Albert the Valiant (*Albrechts-Orden*), Kingdom of Saxony; (ninth left) Order of Honour of the Duchy of Schaumburg-Lippe. It is regretted that space does not permit the full identification on each medal on this and the other arrays that are shown on this spread.

2 An array of medals awarded to an NCO or Other Rank, World War I period. The array begins with two German decorations—the Iron Cross, Second Class, and the Active War Service Cross (*Ehrenkreuz des Weltkrieges*)—and includes Austrian and Bulgarian medals as well as (fifth left) the World War Commemorative Medal (*Weltkriegserinnerungsmedaille*) of the Kingdom of Hungary.

3 An array of medals awarded to a German Officer who served in both the Imperial Army and the forces of the Third Reich (*Wehrmacht*). They include: (left) Iron Cross, Second Class, 1914; (second left) Knight's Cross with Swords of the Royal House of the Hohenzollerns (*Hausorden von Hohenzollern*), Prussia; (third left) Knight's Cross of the Order of the Zähringen Lion (*Orden vom Zähringer Löwe*) with Swords; (sixth and seventh left) Hanseatic Crosses, awarded for service in World War I by the cities of Bremen and Hamburg respectively; (eighth left) Active War Service Cross, Germany; (tenth and eleventh left) Third Reich decorations: Forces Long Service Award in Gold, for 25 years' service, and a similar service award; (twelfth left) Military Service Order, Third Class with War Decoration, dating from the Austro-Hungarian Empire.

4 An array of medals awarded to a German Officer who served in pre-1914 colonial campaigns and in World War I. Among the decorations are (second left) Knight's Cross with Swords of the Royal House Order of the Hohenzollerns, and (third and fourth left) the Royal

Iron Cross, First Class, 1914, displayed on a velvet plaque with the ribbon of the Second Class (the First Class has no ribbon) and a silver laurel spray.

Prussian Order of the Red Eagle and the Order of the Crown: both Fourth Class with Swords and both awarded for service in the South-West Africa campaign.

5-6 Imperial Germany's highest award for gallantry in action – *Pour le Mérite*, often called the "Blue Max". Its French name stems from its establishment by Prince Frederick (later King Frederick I) of Prussia in *c*1667: Frederick was a Francophile and French was spoken at his court. The medal is shown here in its presentation case (6), and as a miniature to be worn in the button-hole on formal civilian dress (5). The decoration was worn suspended from the neck on a black ribbon edged with white stripes interwoven with silver braid. It was not awarded after World War I but was, of course, worn during the Weimar period and under the Third Reich, notably by Reichsmarshall Herman Göring, whose "Blue Max" was won as a fighter pilot on the Western Front during World War I.

7 Qualification Badge of an Observer Officer (*Beobachtungsoffizier*), Imperial German Army.

8 Frederick Augustus Cross (*Friedrich Augustkreuz*), First Class, 1914; Grand Duchy of Oldenburg. See also (6), *page 76*.

9 Commander's Cross with Swords of the Order of the Star of Romania, post-World War I; award of the Kingdom of Romania.

10 Star of the Order of Military Merit, Third Class; Kingdom of Spain. The white arms of the cross show that this award was made for service in peacetime; for war service awards, the arms are red.

11 An array of awards for World War II; Republic of Finland.

12 Decoration for Service in the Red Cross (*Ehrenzeichen für Verdienste um das Rote Kreuz*), Cross of Honour of the Second Class with War Decoration, World War I; Austro-Hungarian Empire.

13 National Defence Cross of 1940; Kingdom of Hungary.

14 World War Commemorative Medal (*Weltkriegserinnerungsmedaille*), World War I; Kingdom of Hungary.

15 Military Cross of Merit (*Militär-Verdienstorden*), Third Class with War Decoration; medal of the Austro-Hungarian Empire.

16 Silver Medal for Bravery (*Grosse Silberne Tapferkeitsmedaille*); Austro-Hungarian Empire.

17 Cross of Liberty, Third Class with Swords; Republic of Finland.

18-19 War Service Cross 1914-1918, in Silver with Sabres (18) and in Gold with Sabres (19); Ottoman Empire (Turkey). This decoration was awarded to both Turkish officers and their German allies (*Imtiaz-Medaille*) for service in World War I.

20-21 Qualification Badges of the Imperial Austrian Air Service, World War I: (20) Military Pilot (*Feldpilot*); (21) Badge for Flying Personnel (*Luftfahrer*).

22 Medal for Valour, World War II; Union of Soviet Socialist Republics. Awarded for acts of bravery to Officers and Other Ranks alike, this decoration has more than a token value: holders receive a small monthly pension and are entitled to travel free on public transport.

Military Headgear

Margaret E. Nobbs

Military headgear has three major functions: distinction, recognition and protection. Up to the turn of the century, the first two functions predominated. Whatever might be the current military fashion, the headgear marked the distinction between the foot soldier, the cavalryman and the gunner—and this tradition is still reflected in modern head-dress, whether it be the full dress cap, side cap or beret.

Although head-dress differed between one nation and another, the fact that certain nations led the general fashion tended to lead to similarities. In head-dress, Russian and Prussian influence was perhaps strongest in the 19th Century; French influence was strong in the period *c*1820-1860, but after the French defeat at the hands of the Prussians in 1870-71, German influence prevailed.

Headgear differed not only between nations but also between arms of service. So far as cavalry was concerned, heavy cavalry generally had metal helmets and light cavalry—hussars and lancers—busbies, shakos or tschapkas (czapkas; see (2), *pages 98-99*). Such distinctions were common to most armies of the time.

By 1900, full dress headgear on the battlefield was becoming a thing of the past. Britain, with its many colonial campaigns of the later 19th Century, was among the earlier nations to adopt service dress, in 1903, introducing the peaked service cap in khaki for field service and reserving full dress for parades. Germany adopted service dress in 1907. The world's armies were evolving into modern fighting machines, and when World War I began in 1914 all the armies concerned had a field service dress—with one exception. The French still clung to the old traditions, with infantry in red and blue and cavalry in cuirasses. The cavalry at least placed hessian covers on their breastplates and helmets, but in 1914 the French infantry advanced and were cut down in their bright red and blue uniforms, which were not replaced by the more sober "Horizon Blue" service dress until the War's second year.

Following World War I, many nations adopted their field service headgear as dress head-dress: the peaked cap, formerly the "walking-out" cap, became the full dress pattern. But national variations persisted; the French, for instance, still clung to the kepi (see examples on *pages 96-97*).

Most collectors of headgear will find attractive one particular national speciality—and for many it will be the German spiked helmet known as the *Pickelhaube* (see *pages 98-99*). First introduced in the 1840s, the *Pickelhaube* was worn in the 19th Century by the Russians, Prussians, British and even the Americans; the French alone of the great powers did not adopt it. By 1900, however, it was actively worn only by the German Army and by some Scandinavian services. Even during World War I, when the leather used for body of the *Pickelhaube* was in short supply, the Germans went to some lengths to preserve their national style: the basic shape was at first stamped out of thin metal and, when this was urgently needed for more serious war purposes, made out of pressed felt. Similarly, even with the introduction of the more practical "Horizon Blue" uniform in 1916, the French preserved the kepi. The British, it is true, did not wear their full dress headgear as service dress—but since the field service cap was of no use for protection, this cannot be said to have been an advantage.

STEEL HELMETS

By 1916, the steel helmet was in general use for battlefield wear; again, with different styles for different armies (see *pages 102-103, 104-105*). The British adopted a domed helmet with a brim all the way round, resembling the traditional pikeman's helmet, while the Germans opted for a helmet with a small peak at the front and a deep "curtain" at the back for neck protection. The French helmet had a small peak at front and back and a crest (or comb) on the top. These designs had not greatly changed by the beginning of World War II: the French helmet had not changed at all; the British design had been very slightly modified; the German helmet had become less bulky and heavy.

Left: *US services' headgear worn by "top brass", May 1945. In seated row, note steel helmet of General Patton (second left), "modified" and "unmodified" peaked caps of General Spaatz, USAAF (third left) and General Eisenhower (centre) respectively, and overseas, or garrison, cap of General Simpson (left). Note also the metal stars denoting rank, worn on the front of the steel helmet or on the left side of the overseas cap.*

Right: *The French Casque Adrian steel helmet, worn here by men of the 3rd Infantry Regiment during World War I, remained in service in World War II.*

Below: *Lightweight "chino" flying helmets are worn by pilots of US 2nd Marine Air Wing based on Okinawa, April 1945.*

Among other types of military headgear worn in the present century—the infinite varieties of peaked caps, forage caps and side caps, and "pill-box" caps—two "hot-weather" types may be given special mention. The British in India adopted during the 19th Century the Helmet, Foreign Service, generally known as the sun helmet or topee, and other countries followed suit. French soldiers in Africa, South America and the Pacific adopted the tropical helmet (with the exception of the Foreign Legion, which retained the kepi with the famous white cover and neck curtain); so also did the Germans in their African colonies. In modern British service, the white tropical helmet remains in service as the full dress wear of the Royal Marines.

Also worn for tropical service was the "bush hat" (see *Insets, pages 90-91, 184-185*). The German Army adopted this for the Pekin Relief Expedition in 1900-01, and at about the same time it was worn by British troops in South Africa. It was later to become the hallmark of Australian troops (New Zealand soldiers also wear it, but blocked in a form similar to the US Army's Campaign Hat), and its widest use in British service today is in the Brigade of Gurkhas.

Various types of utilitarian headgear have been tried out for everyday wear by modern services, the most universal being the beret (see *pages 90-91*). However, in recent years there has been something of a movement towards the more colourful headgear of earlier years: in the British Army, the coloured side cap (see *pages 88-89*) has made a return, while some regiments, particularly the cavalry, have returned to the peaked cap (see *pages 86-87*) of pre-1914 style for walking-out dress. The same trend may be seen to have taken place in the services of other nations.

CARE AND PRESERVATION

Collectors will most often acquire headgear specimens from army surplus stores, arms and armour fairs, and general collectors' fairs. Hats bought from army surplus stores will generally be in good condition but without badges; at arms and armour fairs, where specialists congregate, they will generally have badges and be well-identified—but expensive; at general fairs, junk shops and flea markets, hats may sometimes be bought cheaply, but they will often be in very poor condition.

When a hat or cap is acquired, be sure to check the inside for the name of the maker, the date of manufacture, and the name of the former owner: the first two may be stamped, the last written, on the sweat band. Give the hat a good brush, inside and out, spray it with moth repellent and air it thoroughly to let the spray dry. Polish the peak, if it has one. Marks on peaks may be disguised with shoe polish of the appropriate colour, well rubbed in with a smooth, damp cloth and then polished with a clean, soft and dry cloth.

Clean the badge (if any) with polish—unless it is "Staybrite"—and then scrub it with soapy water, dry well, and coat it with clear nail varnish. Store the hat in a clear plastic bag (these can be bought in many sizes), tucking the surplus plastic inside or tying it with a wire tag. This will keep the hat free from dust, while still allowing it to be seen; the badge, if treated as described above, will stay clean.

Field service caps, also called side caps or forage caps, are best stored in sweater-size plastic bags: the surplus plastic can be tucked inside to give a little packing and help the cap to sit properly on a shelf for display. These caps may also be displayed in a picture frame. Clean the glass well, inside and out, and cut a cardboard sheet of a neutral colour to fit the frame. Fix the caps to this by means of clear nylon thread around the buttons and by sewing the curtain of the cap to the backing sheet. Be sure that the back of the frame fits well enough to keep dust away from the caps.

Finally, a word of warning on display. If the collection is to be displayed outdoors, perhaps as part of a militaria fair, the plastic bags must be removed. If they are exposed to sunshine condensation will occur in the bags and, if they are not properly dried out, mould will grow on the cloth and green verdigris will appear on metal badges and metal chin-strap buttons.

Dress Caps: British Army, Royal Marines, Royal Air Force

1

2

3

4

5

6

1 Dress cap of the 11th Hussars (Prince Albert's Own); this example was made in 1962 by H. Berwald & Co, London. The colour reflects the crimson pantaloons worn by the 11th Hussars, which gave them the nickname of "Cherry Pickers" (or sometimes "Cherubim" or "Cherry Bums"!). The badge is the crest of the Prince Consort, Albert of Saxe-Coburg-Gotha: the regiment received this honour for acting as the escort to Prince Albert when he travelled to marry Queen Victoria. The black leather chin-strap is secured on either side by two plain brass buttons.

2 Cap of a Group Captain, Royal Air Force, made by Gieves Ltd, Old Bond Street, London (1947-1969). Rank is denoted by the single row of gold-wire embroidery on the leather peak. The heraldic crown and wings of the badge are embroidered, with the eagle in gilt; the whole mounted on a mohair band. The patent-leather chin strap is secured by two buttons covered in cloth of RAF blue.

3 Cap of the Royal Scots Greys (2nd Dragoons); the distinctive zigzag "Vandyke band" is a notable feature of this specimen. The body is of blue cloth, the peak is leather, and the plastic chin-strap is held in place by two regimental-pattern buttons. The badge – this example anodised ("Staybrite") – commemorates the capture of the Eagle standard of the French 45th Regiment at Waterloo, 1815, by Sergeant Charles Ewart.

4 Standard pattern cap of the Royal Marines, with non-removable white plastic top. The peak of this modern example, made by Tower Clothiers, is black leather and the chin-strap black plastic. The badge, with Queen's crown, is brass.

5 Cap of the Royal Army Veterinary Corps, which today, with few horses in the British Army, is based at the Remount Depot, Melton Mowbray, where it also cares for the Army's dogs. This style of cap, for wear with battle and service dress, is now worn only by such still-mounted units as the RAVC, King's Troop, Royal Horse Artillery, and the Household Cavalry.

6 The all-scarlet cap – the colour was a regimental tradition – of the South Nottinghamshire Hussars, a Yeomanry regiment, made by J. Compton Sons and Webb (mark "J.C.S. & W Ltd."), with Hobsons, one of the most prolific makers of uniform caps. It has a black plastic-covered peak and chin-strap, held by two brass buttons. The white "Staybrite" badge shows an oak sprig and acorn.

7 Cap of a Sergeant, Irish Guards; rank is denoted by the three bands of brass on the peak. The body of the cap is blue, the band green: other Foot Guard regiments wear the same style of cap, the Grenadiers having a red band, the Coldstreams white (see 8), the Welsh Guards dark green, and the Scots Guards diced red-and-white. The badge, in "Staybrite", is the Star of the Order of St Patrick.

8 Cap of a Sergeant, Coldstream Guards; again with rank denoted by one broad and two narrow bands of brass on the peak; a

7

8

9

10

11

12

Lance-Sergeant has one thick and one thin band, a Lance-Corporal or Guardsman one thick band. The leather chin strap is missing from this example. Note the difference in peak shapes between this cap and (7): the caps were of the same type when issued, but have been shaped according to personal preference.

(Top) Standard pattern cap of a Royal Navy Officer below the rank of Captain, with permanent white cover (formerly worn only on foreign stations) and wire-embroidered badge on a mohair band. (Bottom) Standard cold climate cap for Royal Navy rating, with tally band of HMS Belfast, *the famous cruiser preserved as a museum ship on the Thames.*

9 Cap of the Royal Army Medical Corps; the standard pattern, but with the maroon band and piping peculiar to the RAMC. Made by the Army & Navy Hat & Cap Co. Ltd., 1961, it has "Staybrite" chin-strap buttons and badge.

10 Although this cap of the Royal Army Service Corps (since 1965, the Royal Corps of Transport) lacks its chin-strap and securing buttons, it shows the distinctive white piping, the facing colour of the Corps. Made by Berwald (see 1), this cap is dated 1963; its badge, a King George VI type, is therefore incorrect. With the change of title in 1965, the badge underwent minor changes, the Corps' title being shown on a scroll only.

11 Cap of a Warrant Officer, Class II, of The Blues and Royals (Royal Horse Guards and 1st Dragoons); rank is shown by the five lines of gold braid stitched to the peak. For this rank, the badge is in gilt brass; it is normally bronzed.

12 Cap of the 16th Lancers; in rather poor condition since it is nearly 50 years old. This is an example of the earlier style of "walking out" cap, with a much lower crown and flatter peak than later types. The patent leather peak has "bubbled" with age and the chin-strap, of the same material, has likewise suffered. It is in the distinctive scarlet of the regiment—the only Lancers to wear scarlet tunics with blue facing—and the blue quartered piping at front, back and sides, taken from the cording on the full dress Lancer cap, is typical of Lancer regiments.

Field Service Caps: British Army and Royal Air Force

1

2

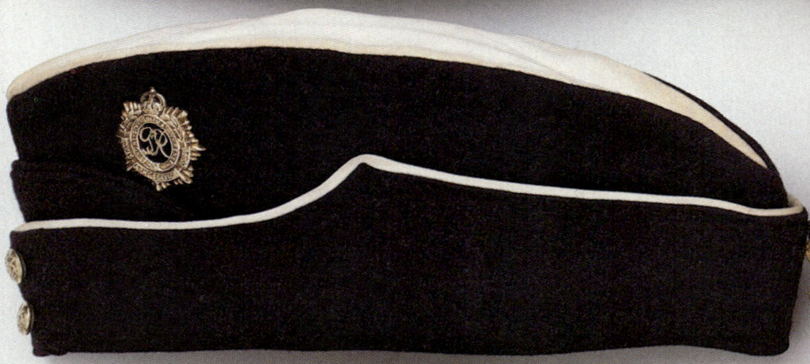

3

4

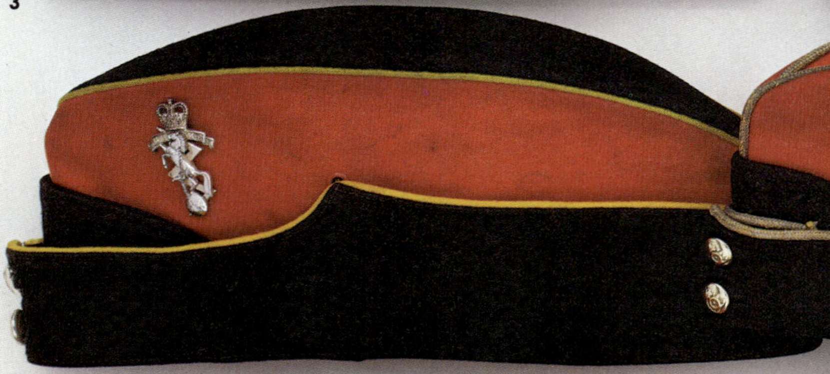

5

6

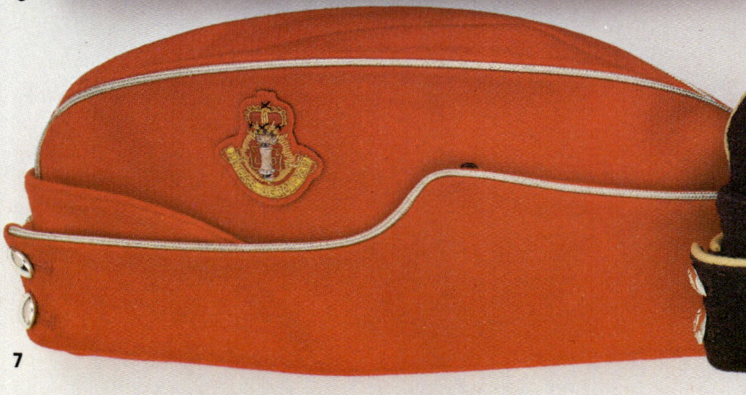

7

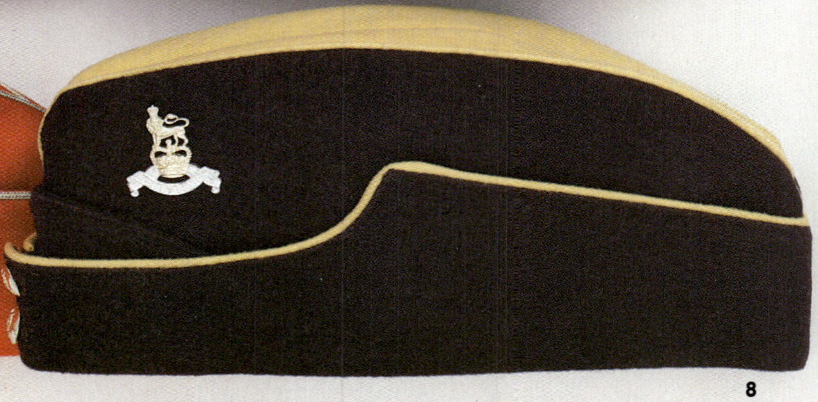

8

Introduced into the British Army in 1894, the field service cap (often called the side cap or forage cap) comprised a body, peak, curtain and chin fastening (for use when folded down). Similar to the Austrian-style cap (and also called by that name) it was originally designed to protect the wearer against cold: the two buttons at the front could be undone to let down the peak and the combined neck and ear flaps, a rather awkward arrangement. Originally issued in blue, it was revived in khaki after World War I, giving way in turn to the beret during World War II.

1 The Devonshire Regiment; a typical example of the coloured field service or side cap. In conformity with the British Army's *Dress Regulations,* it is 4·5in (114mm) high and not less than 3·75in (95mm) across the top. The regulation position for the badge was on the left side, 3·5in (89mm) from the front and 1in (25mm) from the top of the cap. The badge on this example is bi-metal.

2 4th/7th Royal Dragoon Guards; an example of the "Torrin cap" originally worn by the Foot Guards as early as the Crimean War, 1854-56. In this type, the sides are simply pulled down to protect the ears. The badge (missing from this specimen) should again be on the left side of the cap.

3 Royal Army Service Corps; a navy blue cap with white crown and piping in the facing colour of the Corps (since 1965, the Royal Corps of Transport). The badge and buttons are of brass.

4 Auxiliary Territorial Service (ATS) Officer's cap, with red top, blue body, green piping and a brass badge Founded in September 1938 and accorded full military status in 1941, the ATS became the Women's Royal Army Corps (WRAC) in 1949.

5 Royal Electrical and Mechanical Engineers (REME); in the colours, REME combines the facing colours of the Royal Artillery and Royal Engineers: blue, red and yellow.

(Top) 9/12th Royal Lancers (Prince of Wales's); this is the former 12th Lancers' cap of the type defined in Dress Regulations of 1900: *the 12th were the only Lancers to wear scarlet. (Bottom) Royal Corps of Signals; defined in* Dress Regulations of 1934.

9

10

11

12

13

14

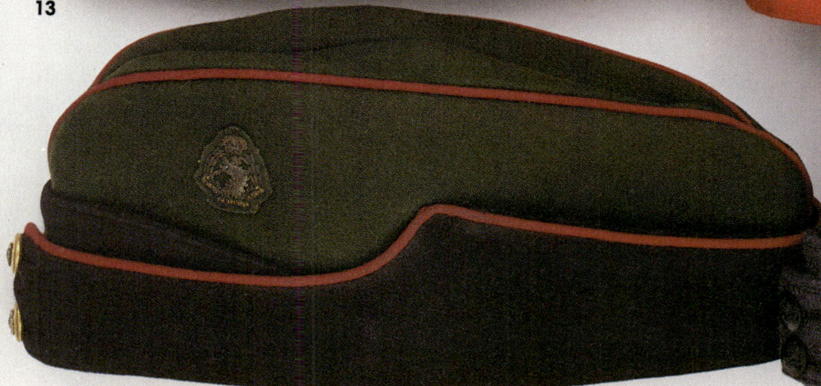

15

16

6 Royal Artillery Officer's side cap, in the distinctive colours of the Gunners and with a wire-embroidered badge showing the grenade and the motto *Ubique* ("Everywhere"). The edges are piped in gold cord.

7 Leicestershire and Derbyshire (Prince Albert's Own) Yeomanry; a scarlet side cap trimmed with silver cord, with a gold and silver wire-embroidered badge.

8 Royal Army Pay Corps; standard pattern cap with yellow crown and piping; "Staybrite" badge.

9 Royal Army Medical Corps; the crown and sides are in dull cherry and the gold cord piping denotes commissioned rank. The badge is gilt and silver; the button gilt.

10 Middlesex Regiment; the standard blue cap with top and peak in the regiment's lemon-yellow facing colour. The badge is silver, as laid down in *Dress Regulations,* and the buttons gilt.

11 Royal Anglian Regiment; a modern style cap in the blue and red typical of Royal regiments, the only distinctive factors being the badge and buttons. The "Staybrite" badge, featuring the Castle and Key of Gibraltar within the star, and buttons contain elements from the former insignia of all the regiments that have made up this modern amalgamation since 1964.

12 3rd/6h Dragoon Guards (Carabiniers); the side cap of an amalgamation of 1922, combining the blue of the tunic of the 6th (the only Dragoons or Dragoon Guards to wear this colour) and the yellow facing of the 3rd. The badge similarly combines the Prince of Wales's Feathers of the 3rd and the crossed carbines of the 6th. In 1971 the regiment was amalgamated with the Royal Scots Greys (2nd Dragoons) to form the Royal Scots Dragoon Guards (Carabiniers and Greys).

13 Royal Gloucestershire Hussars; khaki service cap of the standard type, but distinguished by the regimental badge and buttons. The badge shows a portcullis with chains with a ducal coronet above. These are the insignia of the Duke of Beaufort, and the regiment's full dress uniform was in a shade known as "Beaufort Blue".

14 Leicestershire and Derbyshire (Prince Albert's Own) Yeomanry; an Officer's cap, in scarlet cloth edged with silver cord, with "Staybrite" badge and buttons. Prior to amalgamation in 1957, both the Leicestershire Yeomanry and Derbyshire Yeomanry wore scarlet caps, the former with silver piping and the latter with gold: the senior regiment's cord was adopted on amalgamation.

15 Royal Army Dental Corps; a fairly rare cap from a small Corps, with green body and sides, red piping and a blue bottom – the combination of blue and red denoting "Royal". The badge is wire-embroidered and the buttons gilt.

16 Royal Air Force; a serge side hat of the kind worn by every British airman during World War II. Like its Army counterpart, it is made of the same material as the battledress of the period. This example has a plastic badge and buttons.

Berets: British Army, Royal Marines, RAF, Women's Services

The beret is now the undress headgear of many armies, navies and air forces all over the world. The design was influenced by the need for a suitable headdress for tank crews. When the British Army's Royal Tank Corps (later the Royal Tank Regiment) adopted full dress for the first time in 1922, they opted for the beret as full dress headgear; it is still worn as such by the regimental band. Note that after issue, berets tend to be shrunk, kneaded and rolled into a shape that suits the unit's or wearer's preference; this explains the variations in shape and condition seen in the berets shown here.

1 Royal Marines; the hard-won green beret is now a distinctive emblem of the Royal Marines, but during World War II it was worn also by other Commando units. This example has a black-painted badge.

2 Parachute Regiment; the famous "Red Beret" of "the Paras" is a prime example of the introduction of distinctive colours for berets in place of the standard khaki (or, later, blue-black). This rather badly-worn example has a modern "Staybrite" badge.

3 Special Air Service (SAS); a modern example of another beret of a distinctive colour. This an Officer's beret in near-perfect state, having been subjected to no "modification", with an embroidered badge.

4 Welsh Guards; standard khaki beret, with worsted badge sewn to it.

5 Leicestershire Regiment; a standard issue beret given regimental individuality by the placing of a green cloth patch behind the badge (a similar practice was adopted with the much-disliked "Broderick cap" of 1903). The bi-metal badge is of pre-1946 pattern: in that year the regiment was granted the "Royal" title and the badge was restruck to accommodate this on the scroll.

6 Royal Regiment of Fusiliers; the red-over-white hackle was, from 1892 onward, a distinctive emblem of the Royal Northumberland Fusiliers, the senior regiment of this 1968 amalgamation which also included the Royal Warwickshire Fusiliers, Royal Fusiliers and Lancashire Fusiliers.

7 Royal Air Force; Officer's issue beret in RAF blue, dated 1957, with a two-piece badge of gilt metal.

8 Royal Anglian Regiment; beret of an amalgamation dating from 1964 and taking in many of the former Anglian County regiments. The badge incorporates various insignia from the original regiments involved.

9 British Army; World War II (1943) gabardine beret, issued with fatigue battledress in the same material. The badge is plastic: these economy wartime badges were much despised by those to whom they were issued, but many types are now rare and are highly collectable.

10 Lovat Scouts; the tam-o'-shanter with the black-and-white toorie (pompom) is typical of the undress headgear of some Scottish regiments. This example has a black-and-white diced band. The badge is white metal. The pull ties

The Bush Hat is closely associated with Australian troops and Gurkhas, but it was worn by other servicemen: this a Royal Air Force example, worn in the Far East, World War II and later.

which run inside the head-band and are visible at the back are cut in the traditional "swallow tail".

11 11th Hussars (Prince Albert's Own); this brown beret with a dark maroon band is of a type peculiar to the regiment and represents one of the earliest deviations from the standard pattern. The badge is wire-embroidered.

12 Army Air Corps; the light-blue beret is worn by all ranks of the Corps. The "Staybrite" badge, of the pattern adpoted on the Corps' re-formation in 1957, is backed by a dark blue-black square. Men of other units seconded to the Corps wear the same beret but with their own unit's badge.

13 Auxiliary Territorial Service (ATS); dated 1942, this is an example of the original headgear introduced during World War II. The "floppy-top" hat was much disliked by the women who wore it; ATS Driving Instructors wore the chinstrap over the crown as an unofficial mark of distinction.

14 Women's Royal Air Force (WRAF); something of a rarity, since it is the cap of a Group Captain, a rank few women attain. It is adapted from the standard pattern, with a smaller crown; note the fine wire-embroidered badge.

15 Women's Royal Army Corps (WRAC); dark green cap of an Officer, distinguished by the badge in gilt metal and silver. The black plastic strap is secured by two "Staybrite" buttons. Other Ranks wear a cap of the same style, sometimes in pressed felt, with a badge of gilt metal and white metal.

Headgear of the US Army, Navy and Women's Services

Like most aspects of the armed forces, the development of uniform, however modern, is somehow dictated by tradition. The American services have a long tradition of French influence in uniforms; but as uniform fashions change, the world is now following the USA. This trend is most noticeable in headgear; American influence can be seen worldwide.

1 US Army Cold Weather Cap. This style of cap, originating in headgear worn by the Soviet forces, was generally issued for service in the Korean War, although it was in use earlier. The construction is simple and the use obvious: the ear flaps and peak are worn down in cold weather and up, as shown, in temperate conditions. Rank or other insignia were often worn on the front of the upturned peak.

2 US Army Knitted Cap of World War II, with a peak strengthened with laminated card. Originally designed to be worn under the steel helmet, this was the US equivalent of the British "Cap, Comforter" worn by Commandos and other forces; General Patton, it is said, hated this style and even forbade its wear unless hidden beneath the helmet. This ruling was largely ignored by the men of his US 3rd Army—except of course, in areas where Patton might appear in person.

3 US Army; an adaptation of the knitted cap seen at (2); this is known as the "jeep cap" or, thanks to the power of television, as the "Radar cap", from the character who wore it in the movie *M*A*S*H* and the subsequent TV series.

US Army Officer's Cap of 1920 Pattern, with the badge of an Aviation Cadet of the Army Air Corps, the designation of the service from 1926 until 1941.

Although unofficial, it was popular; it represents one of the numerous examples of commercial copies of standard issue items.

4 Unidentified fatigue cap, thought to be of US origin and similar to the knitted caps shown at (2) and (3). Although of the same shape and design as (2), it has a chin strap in the same material.

5 US Navy; Enlisted Man's Summer Issue Fatigue Cap, worn with white summer uniform. This style typifies the US Navy "gob", or rating, and was made famous by the cartoon character "Popeye the Sailor". In wear, it is often distorted into wierd shapes (or total shapelessness). First introduced after World War I, it is still current issue.

6 US Women's Army Corps; Enlisted Women's Peaked Cap, similar in

style to the headdress now worn by other women's services internationally. On this example, the badge, instead of being mounted on a brass disc, is mounted within a circle of the same metal.

7 Women's Army Corps; Full Dress Cap for wear by an Officer with dress blue uniform, with a gold lace band on the uniform blue body of the cap. In pattern it is the same as (6): the peak, as in (6), has rows of multi-stitching to give it firmness. The badge, with neither brass disc nor circle, is the standard pattern.

8 US Army Nurse Corps. This peaked cap of World War II vintage is similar in many respects to the British Auxiliary Territorial Service (ATS) cap shown at (13) on *pages 90-91*. It has a floppy top, a chin-strap/band in the same colour and material, and a stiffened peak. All US Army Nurses were commissioned officers, and the badge is the standard bronzed issue. A point of interest is the manner in which the badge was worn: although this is not apparent in the photograph, it was placed off-centre so that when the cap was worn, the badge was positioned over the left eye. This is a rare item: few of these hats survived World War II.

9 US Army; Enlisted Man's Peaked Cap, World War II, winter issue.

10 US Army; Officer's Full Dress Cap, the current issue for wear with the dress blue uniform. It has a black leather peak and a gold lace chin-strap secured by two standard US General Service gilt buttons. The badge is larger than that worn by an Enlisted Man (see 9) and has no backing disc.

11 US Army; Officer's Cap, World War II, summer issue. Widely worn in the Pacific Theater, this style is perhaps typified by the famous cap of General MacArthur.

12 US Army; Officer's Cap, World War II, winter issue, with bronzed badge and brown leather peak and chin-strap. The badge is slightly smaller than that shown at (11). This style of cap, especially when worn by aircrew of the US Army Air Force, was often "customised" by the removal of the wire or rubber stiffening grommet from within the rim, allowing it to be turned down at the sides to produce what was termed the "50 Mission Crush". The intention was to make it appear that the hat had been worn on 50 flying missions – the look being the result of wearing radio earphones over the cap during those missions. Such harmless vanities are common to most armed services.

13 US Army; Herringbone Twill (HBT) Fatigue Cap, designed in 1941 for wear with the HBT fatigue suit and a popular item of headgear among GIs. Again, as in (6) and (7), it has a multi-stitched peak for stiffness.

14 US Army; HBT Fatigue Cap of the type known as the "Daisy Mae" after the character in the famous "Li'l Abner" comic strip. It was originally issued in blue denim for wear with blue denim fatigues, and later in khaki and olive drab, and was designed to replace the wide-brimmed Campaign Hat worn since before World War I.

Overseas (Garrison) Caps: US Army, Navy, Women's Services

1 Overseas Cap of Enlisted Man, US Army. Worn in all Theaters of War from 1918 to the present, it is perhaps the longest-serving US uniform item. Also known as Garrison Caps or Field Caps, they were issued to all ranks; in khaki cotton or worsted for summer wear and olive-drab wool for winter wear. They were habitually–but unofficially–worn with the crown forced down to produce pronounced peaks fore and aft. The example illustrated is a winter issue cap of World War II.

2 Rank chevron of a Private First Class (PFC), US Army; worn on both uniform sleeves–and, of course, inverted, not as shown.

3 Overseas Cap, US Women's Army Corps (WAC), World War II; officer's summer issue. All officers' overseas caps, for both men and women, have gold and black piping as seen.

4 Overseas Cap, WAC, World War II; officer's winter issue. This cap was designed to replace the Kepi-style peaked cap which was first issued.

5 Overseas Cap, WAC, World War II; enlisted personnel, summer issue. The WAC had branch of service piping in old gold and moss-green. All other branches of service, US Army, had their own piping for Overseas Caps: Infantry, pale blue; Artillery, scarlet; Cavalry, yellow; Air Corps, gold and deep blue; Signal Corps, white and red; Military Police, green and yellow; Armor, white and green; Ordnance Corps, yellow and brown; Tank Destroyer, yellow

Bandswoman's Cap, Central Band (disbanded 1970s), Women's Royal Air Force, Britain. It has a gold-wire embroidered badge and a strip of gold tape sewn behind the front turn-up.

and green; Medical Corps purple and white; Engineers, white and red; Quartermaster Corps buff.

6 Rank insignia, Petty Officer Boilerman 1st Class, US Navy; winter issue.

7 Overseas Cap, US Army, World War II; officer's winter issue, with rank insignia of Captain on left side.

8 Rank insignia, Petty Officer 1st Class, Lithographer, US Navy; worn on sleeve of summer white uniform.

9 Cap Badge, US Army, officer; worn on peaked caps by officers of all ranks. Enlisted personnel wore a smaller version of the badge, backed by a brass disc.

10 Overseas Cap, US Army, World War II; summer issue with rank insignia of Major-General.

11 Rank insignia, Enlisted Aircrew, US Navy; worn on white summer uniform.

12 Overseas Cap, US Army, World War II; officer's cap of heavy chocolate-brown gabardine. Since officers purchased their dress uniforms privately from military tailors, a wide variation in styles and materials of their uniforms and headgear may be found.

13 Overseas Cap, US Army, current style for officers. The US Army changed its uniform colour from a khaki shade to the green of this cap – officially Army Green 44 (AG44) – in the late 1950s. This example is the cap of an Airborne officer, with the Airborne patch on the right; enlisted men wear the patch on the left.

14 Collar insignia, enlisted personnel, US Army; worn on the right lapel of the uniform. On the left lapel is worn the branch of service insignia, on a brass disc: Infantry, crossed rifles; Artillery, crossed cannon; Military Police, crossed flintlock pistols; Ordnance, flaming grenade; Signals, crossed semaphore flags; Cavalry, crossed sabres; Engineers; castle; Quartermaster Corps, US Eagle atop a wheel, enclosing crossed sabre and key.

15 Rank chevron, Corporal, US Army, World War II: again, unfortunately, shown inverted.

16 Overseas Cap, US Army; enlisted personnel winter issue, with the enamelled badge of 2nd Armored Division.

17 Overseas Cap, US Army; current issue, enlisted personnel.

18 Overseas Cap, US Civil Defense, World War II; with enamelled badge of the organisation. Although the United States suffered no major bombing attacks in World War II, there was in existence a large Civil Defense organisation of volunteer Air Raid Wardens, Fire Guards and Rescue Squads. They took part in regular exercises, especially in the East and West Coast areas, where any major sea or air attack could be expected.

19 Overseas Cap, US Marine Corps; enlisted man's cap with the Globe and Anchor insignia of the Corps.

20 Rank chevron, Staff Sergeant, US Army, World War II; winter issue item.

21 Rank chevron, Sergeant Major, US Marine Corps.

1 Kepi of the French Foreign Legion (*Légion Etrangère*). This pattern altered very little from the 1870s, the time of the Franco-Prussian War, until the middle of World War I. It has a sunken top and is made from the standard uniform material of heavy red and blue woollen cloth, with blue braid quartering. The grenade insignia of the Legion, in red cloth, is sewn to the broad blue band. It is ventilated by two "pepper-pot" vents on either side, and the leather chin-strap (missing from the example shown) was attached by two hollow-backed brass buttons with Legion insignia.

2 Kepi of similar pattern to (1), dating from the same period and worn by line infantry regiments; in this case, the 79th Infantry. At the beginning of World War I, before the general issue of protective helmets, a steel skull-cap was worn beneath the kepi; with the introduction of "Horizon Blue" uniform later in World War I, covers of the same colour were worn over the kepi.

3 Kepi of a Captain, Foreign Legion, *c*1920. The regiment is shown by the gilt metal badge; the rank by the three parallel bands of gold-wire lace, with two further bands descending from crown to band at front, sides and back. The kepi of a Lieutenant had two gold bands; a Sous-Lieutenant, one.

4 Kepi of a General of Brigade; this pattern was worn during World War I and into the 1920s. Rank is shown by the broad band of gold-wire embroidered oak leaves on the blue background; the crown bears a quatrefoil of four lines of gold-wire braid, weaving intricately over and under at the intersections, and three more lines of braid descend vertically from the crown to the band at front, back and sides. Note that the quatrefoils on the kepis of lower commissioned ranks are less elaborate. In 1916, when the "Horizon Blue" uniform was introduced for the Metropolitan Army, a plain active service kepi was introduced in place of the one shown; its only decoration being two metal stars on the band and oak-leaf embroidery on the black velvet chin-strap.

5 Kepi of a Major, Foreign Legion; this style was worn from the late 1930s until 1946. The indication of rank follows the system described at (3); in this case, with four bands of gold braid around the kepi and three bands descending at the quarters. A Lieutenant-Colonel had five rows of braid, three gold alternating with two silver (*vice versa* for Cavalry; it should be noted that all French Cavalry units have the distinction of silver braid and buttons), and a Colonel five rows in gold or silver, depending on arm of service. Note also that the wire-embroidered grenade insignia is larger and more elaborate than on the later version shown at (12).

6 Kepi of a General of Division, World War II period. This is similar to the kepi of a General of World War I, shown at (4), except that instead of one broad band of oak-leaf embroidery there are two narrower ones. Note also that the height of the crown is not so great.

7-9 Foreign Legion; three kepis of Legionnaires (Other Ranks), dating from the 1960s. The famous white covers of the Legion have been removed from (7) and (8) to show detail of crown and band. This reveals on (7) a grenade badge in green – definitely non-regulation and something of a mystery. The standard red grenade is seen on (8). During the 1980s, the removable white cover was discarded: the present-day kepi is made with an integral covering of white cotton duck or, sometimes, plastic or PVC material.

10 Foreign Legion; kepi of an Other Rank, Legion Cavalry, dating from the 1970s; note the silver buttons distinguishing arm-of-service, as explained at (5). At this period, the kepi ceased to be made in the traditional red and blue woollen material: this headdress of the 1970s is of khaki chino-type material – over which, as seen, the removable white cover is worn.

11 Kepi of a General of Brigade; the style worn from after World War II until the present day. It is similar in style to (4), with a single band of oak leaves, although the blue band has been increased in area at the expense of the red crown; because of this, the gold-wire embroidery is more widely spaced.

12 Foreign Legion; kepi of a Major; the type worn from the 1960s to the present day. The gold lacing above the band and on the crown, distinguishing rank, is the same as on the earlier kepi shown at (5). This officer's kepi is still made of the traditional red and blue material, but the extent of the blue band has much increased. The kepi itself is more blocked and cylindrical in shape. Comparison between (5) and the example shown here shows how shape and style have changed since World War II, while the method of distinguishing rank has remained the same.

Green berets of the French Foreign Legion; 1980s issue style, in which nylon or other synthetic edging has replaced the leather band with draw tapes. The insignia of the winged hand clutching a dagger is common to all French airborne troops; the flaming grenade on the lower beret is worn by all Legion units (in white metal by Cavalry).

Headgear of Imperial Germany, 1900-1914

During the 1830s, a movement for the reform of military uniforms swept a number of countries, where both uniforms and headgear underwent significant changes. Russia, at that time a leader in military fashion, was experimenting with a new leather helmet and, as rumour has it, King Frederick William of Prussia saw this on the Tsar's desk during a visit and, recognising its advantages, quickly had it copied for the Prussian Army. The new helmet, best known as the *Pickelhaube*, had a conical leather body with peaks at front and back, the front peak edged with metal. On the top was a cruciform ornament with a metal spike in the centre; a metal rib ran from the top ornament to the edge of the back peak. The Prussian Artillery adopted a ball top in place of the spike in 1846; it is interesting to note that in Britain, where this style, in blue cloth, was adopted in 1878, it took three years for the authorities to introduce a ball top as a safety measure for gunners. Although modifications were made in 1857, 1860, 1871, 1887 and 1891, the basic pattern of the Prussian *Pickelhaube* remained virtually unchanged except for height and shape. In 1916, however, after wartime *ersatz* versions in pressed tin and felt had appeared, the *Pickelhaube* was finally abandoned in favour of the utilitarian steel helmet.

1 Shako of an Administrative Official of the East Asian Expeditionary Force sent from Germany to take part in the relief of the Legations at Pekin during the "Boxer Rebellion" of 1900.

2 Tschapka (Czapka) of an Other Rank, Prussian Lancer line regiment; Model 1867. As its name suggests, the style of this Lancer's helmet originated in Poland; it was adopted by the Lancer regiments of several countries and is characterised by the flat, square top. Note that there are Prussian state cockades at either side on this example, and that a field "plume" is worn in place of the full dress plume of horsehair.

3 Dragoon Helmet of an Other Rank, Hesse; 1887 Pattern. In leather with silver fittings, the helmet has a fluted spike and bears the Hessian state badge. The Imperial cockade is mounted on one side; the State cockade on the other.

4 Shako of an Other Rank, Prussian *Jäger* (Rifle) battalions of the line; Model 1892. The body is of black leather with brass mountings; the overlapping black leather top is stitched around the body. Behind the badge, a leather slip provides accommodation for the "plume" or cockade. No cockade is worn on the left side. The leather chinstrap, as is usual in all styles of German Other Ranks' helmets, has twin buckles for adjustment, and the back of the shako is shaped to the head. *Jäger* battalions had worn the *Pickelhaube* until 1854; this style of shako survived as the German Police helmet until as late as the 1950s.

5 Helmet of an Officer, Cuirassier Regiment Tsar Nicholas I of Russia (Brandenburg) No 6. As a line

regiment, this unit should have worn a steel helmet, but because Tsar Nicholas was its Colonel-in-Chief it adopted the style reserved for Guard-Cuirassiers, with a mixture of silver and gilt fittings.

6 Tschapka plate of German silver for an Officer of a Württemberg regiment, *c*1910. After the establishment of the German Empire in 1871, its various states still retained their own badges and often had different uniform features.

7 Helmet plate of an Other Rank in a Saxon reserve regiment, *c*1910; brass and German silver.

8 Helmet of a Prussian General; Model 1891/97. A typical helmet for General Officers, with square peak and fluted spike. The cruciform top is held by stars (one is missing) and the Guard Star plate has an enamelled centre.

9 Helmet of an Other Rank of a Baden Artillery regiment; Model 1871/97. Imperial and state cockades are worn at the sides and, since this is an Artillery helmet, it has a ball top.

10 Shako of the type worn by Other Ranks of Bavarian *Jäger* battalions and Cyclist companies. The evolution of style will be seen when this is compared with the earlier *Jäger* shako at (4): it is notable that the shape has grown squatter.

11 A rare Other Ranks' helmet (possibly for Mountain Troops) of *ersatz* ("replacement") type, dating from 1914-15. It is made of cork covered in khaki cloth; the cockades are riveted to the helmet to the rear of the chin-strap fittings. The plate is that of Prussia.

12 Helmet of an Other Rank, Saxon Infantry; Model 1895/97. The mountings are all brass, with the state plate in bi-metal (brass and German silver).

13 Helmet plate of German silver for an Other Rank of a Prussian Dragoon line regiment; on an Officer's helmet plate, the eagle had a pierced-out crown.

14 Helmet plate of aluminium-bronze, Bavarian Army; this alloy was adopted after 1895 in order to reduce the weight of the headgear of which it formed a part.

One of the most prized pieces of headgear in any collection: a Prussian Garde du Corps, *Imperial Bodyguard, helmet of c1900. This particular example is the helmet worn by Kaiser Wilhelm II himself.*

Headgear of Imperial and Nazi Germany, 1900-1945

1 Cap of an Officer of Marines, Imperial Germany. These troops, although seaborne, received infantry training and had much the same status as the British Royal Marines. The cap, with its distinctive small peak, is of navy-blue cloth with a white band and white piping. It bears on the front an Imperial cockade but, unlike many of the helmets of the pre-World War I period shown on *pages 98-99,* bears no state cockade, since the Marines were classed as *Reichs* (Imperial) troops under the direct command of the German Emperor.

2 Summer issue cap of an Officer, Imperial Germany Navy; the badge and cockade are wire-embroidered. As in all navies of the period (and the custom is still observed by some), "official" summertime or service in the tropics saw the addition of a white top or cover to the cap. In all countries, naval uniform styles tend to be the most tightly bound to tradition: except for changes in insignia, most have undergone no great alteration during this century. This style of cap endured in German service through two World Wars and is worn today, with slight modification, by the Navy of the German Federal Republic.

3 Cap of a Protestant Chaplain of the German Army, World War I. The band is violet, the colour traditionally worn by chaplains of all services, nations and denominations. This example bears the plain cross of a Protestant chaplain between the Imperial and state cockades; Catholic chaplains wore crosses with three small circles at the terminals.

4 Other Rank's cap of the type known as the *Krätzchen,* 10th Regiment of Hussars, Magdeburg, *c*1900. It has a green crown with yellow piping; the Imperial cockade is worn above the state cockade of Magdeburg.

5 Officer's cap, 17th Brunswick Hussar Regiment No 17; this particular example was the cap worn by the Colonel-in-Chief, Duke Ernest Augustus of Brunswick-Lüneburg. Note the similarity of colour and piping with (4), also a Hussar's cap. A metal regimental badge, the *Totenkopf* ("death's-head"), is worn between the Imperial and Brunswick cockades.

6 Rating's cap, *Reichsmarine,* with the class and name of the light cruiser *Berlin* on the tally band. The badge is that of the Weimar Republic—a black eagle with red feet on a yellow ground—worn from 1919, following the fall of the Empire, until 1933, preceding the establishment of the Third Reich.

7 Cavalry Officer's cap, *Reichswehr* (the Army of the Weimar Republic). Like the Imperial Army caps it bears a state cockade (that of Prussia), but in this case worn above the national cockade, on the band, which comprises the badge of the Weimar Republic within an embroidered wreath of oak leaves. As in (4) and (5), the golden-yellow piping is the *Waffenfarbe* (arm-of-service distinguishing colour) of the Cavalry.

8 Cap of a *Luftwaffe* (Air Force) General, Third Reich, 1933-45. The national emblem (*Hoheitszeichen*)

Black felt bicorne hats of (top) Admiral, Imperial German Navy; (bottom) Admiral, Kriegsmarine, *Navy of the Third Reich. They differ only in insignia on the buttons.*

is mounted on the front of the cap cover and the *deutsche Reichskokarde,* the black, white and red cockade surrounded by oak leaves and flanked by wings, on the black mohair band. For all generals, the piping, cords and insignia were, as here, in gold; for other officers, the decorations were in silver, as seen at (10). NCOs and ORs had piping in *Waffenfarbe,* national emblems and cockades stamped out of aluminium, and black patent-leather chin-straps.

9 Fur cap for winter wear of the *Waffen-SS* ("Armed SS"; military formations serving alongside the Army); made of dyed rabbit-skin and field-grey cloth. The *Hoheitszeichen* and the *Totenkopf* insignia of the SS, in metal, are mounted at the front; the back of the cap could be turned down to protect the wearer's neck.

10 Summer issue cap of an Officer, *Luftwaffe,* with removable white cloth cover. As noted at (8), the silver cords, silver metal *Hoheitszeichen* and silver-embroidered wreath and wings enclosing the cockade (all this embroidered in one piece), denote an Officer below the rank of General. Some Warrant Officers also wore silver cords, but with insignia of aluminium like those of NCOs and Other Ranks.

11 Officer's cap of the *Wehrmacht,* the German Army (although the term could be used to embrace the Navy and Air Force also) of the Third Reich; with the white *Waffenfarbe* of the Infantry. This style was worn by both Officers and NCOs with service dress and undress uniform, and by Officers alone with parade dress when off duty, walking-out and dress uniform.

12 Peaked cap of a General Officer, *Wehrmacht;* a non-regulation example with silver cords rather than gold (see 8). Such headgear was not uncommon in combat zones, where promotion might be both speedy and frequent and gold cord might not be available to a newly-promoted General.

13 Forage cap for NCOs and Other Ranks of the *Luftwaffe,* Third Reich; with cloth *Hoheitszeichen* and cockade. As may be seen, the strap could be unbuttoned, allowing the side parts of the cap to be turned down as protection against cold, with the strap buttoned beneath the chin.

1 German steel helmet, Model 1916; with traces of camouflage colouring. The helmet marked a breakthrough in protective headgear; it was the German Army's first steel helmet, replacing the leather helmet—the *Pickelhaube*, see *pages 98-99*—which had proved to be inadequate for trench warfare. The example shown is one of the original styles recommended to the Prussian War Ministry in August 1915 by Professor Doctor Bier, Surgeon-General in the German Navy. Troop trials had taken place by the end of the year and the helmet was issued to the troops taking part in the assault on Verdun in January 1916. In the following months a general issue was made: some 7·5 million helmets had been manufactured by the war's end. Following suit, in 1916 most combatant nations introduced metal helmets of various designs: the French already had a steel skull-cap for wear beneath the kepi; but only the introduction of the steel helmet succeeded in reducing the number of fatal head wounds.

German steel helmet, Model 1916, with 0·31in (8mm) steel plate for added frontal protection for observers and snipers in trench warfare; introduced in 1917.

2 Model 1916 steel helmet with emblem of the Weimar period, 1919-33. Stocks of the Model 1916 helmet were used by both the *Reichswehr*, the Weimar Army, and the *Freikorps* (see 4). No insignia were worn from 1919 to 1923, when provincial unit insignia—the one seen here is for Württemberg—were introduced.

3 Model 1916 helmet without peak; a rare example of a design produced for Turkey, Germany's ally, towards the end of World War I. Only 5,400 were made and none was delivered; most were acquired by the *Freikorps* (see 4) in 1919.

4 Model 1916 helmet hand-painted with *Freikorps* (Free Corps) insignia; the chin-strap is not the original. This para-military association of World War I veterans was formed in the interregnum between Imperial Germany and Weimar, to protect the eastern frontier and combat internal subversion; many of its units degenerated into extreme right-wing strong-arm squads, supporting the rise to power of Hitler's Nazi Party.

5 Flying helmet of a Pilot, German Flying Corps, World War I. This finely-made leather helmet is not an issue item but a privately-purchased piece.

6

9

7

10

8

11

6 German steel helmet, Model 1935; with decal of *Waffen-SS*. This was a re-designed, lighter version of the Model 1916 helmet and was worn by the Army, Navy, Air Force, *Waffen-SS*, Police and various special formations. The colour for the Army, Navy, *Waffen-SS* and Police was grey, for Air Force ground troops, grey-blue (see 7); all arms had their own decals. On the example here, there are SS runes on the right side (as seen) and the black, white and red national emblem on the left: the distinguishing insignia of the elite *Leibstandarte Adolf Hitler*. The Model 1916 helmet remained in service alongside the Model 1935 until early in World War II

7 Model 1935 helmet in the blue-grey colour worn by the *Luftwaffe* (Air Force); with the Luftwaffe eagle decal on the left side.

8 German steel helmet, Model 1918; a special pattern, often called the Signals or Artillery helmet, with "ear cut-outs" to allow field telephone headphones to be worn. Few were issued during World War I, but it was later worn by Cavalry of the *Reichswehr* (Weimar Republic Army). Examples are to be found in camouflage paint as well as the standard grey. The helmet seen here belonged to a German Cavalry Officer of World War I; he later joined the *Wehrmacht* (Army of the Third Reich) and, as seen, had its insignia placed on his old helmet. Examples of this form of adaptation are rare.

9 Model 1935 helmet with an unusual turned-up rim, in camouflage colouring; as issued to field troops of the *Luftwaffe*. It bears no insignia: after 1943, no insignia were worn on the steel helmets of German combat units.

10 German Paratrooper's steel helmet; Paratroop units were established by the Army and *Luftwaffe* in 1936, but it was not until 15 March 1938 that special equipment, such as the helmet, was introduced. On 1 January 1939, the Army's Paratroop units were absorbed into the *Luftwaffe*. The rimless helmet, of the kind later adopted by paratroops of all nations, underwent no significant modification throughout the War. Olive-drab helmet covers were issued in late 1940 and green camouflage covers in 1941. From 1943, brown camouflage helmets of the type shown here, with hessian covers with cord or chicken-wire netting, were issued. The Paratroopers of the *Waffen-SS* also wore helmets of the type shown here.

11 German SSK 90 protective steel helmet for aircrew; a rare model, introduced in May 1941, designed to protect flying personnel from small-calibre bullets and shell fragments. The helmet is made from several sheets of chromed steel, overlapping in "lobster tail" fashion; the whole being leather-covered. Note that a rear view is shown: at the front, a raised leather crest (its termination at the crown visible in the photograph) was fitted so that the helmet might more easily be put on or taken off with one hand only.

1 Imperial Japanese Army tank crew helmet, World War II; a rare item for the collector. Tank helmets were intended to protect crewmen's heads from injury on the interior of the tank, rather than from shot and shell, so this crash-helmet style headgear of canvas-covered compressed cork is a practical design of a type common to many modern armies. It has a leather and canvas lining, leather chin straps, and laced openings for earphones at the sides. This is the summer issue helmet; the winter helmet was fitted with a fur-lined leather neck and ear protector. The five-pointed yellow star is a common insignia on Japanese headgear of World War II.

2 Damage Control (DC) steel helmet, worn by British Civil Defence workers, World War II; again, somewhat rare. The standard helmet of its type, fitted with a neck protector of oiled canvas to guard the skin against such corrosives as mustard gas which, it was thought, might be used against the British population in air attack. The "DC" logo has been painted over an original "W" for "(Air Raid) Warden".

3 Infantryman's steel helmet, Italian Army, World War II; a common type. This is the Model 1935 helmet that replaced the earlier French-type issue. It has a leather lining with a cord adjustment for size. These are sometimes found with stencilled arm-of-service insignia applied in black paint.

4 Infantryman's steel helmet, French Army, World War II; not uncommon. Known as the *Casque Adrian* and introduced during World War I, this style of helmet was also produced for Belgium, Romania, Serbia and Russia, from May 1915 onwards. The World-War-II model was made of improved steel. All French helmets bore arm-of-service badges: this one has what is probably the most common insignia, that borne by Infantry, Cavalry, Fortress Units, the Supply Train and the French Foreign Legion.

5 Steel helmet, British Army, World War I; fairly common. The Model 1916 or Mk 1 steel helmet—also called "Brodie's Pattern"—provoked some opposition when first introduced; it was remarked that head wounds increased without taking into account the fact that fatal head wounds diminished. The leather chin-strap is secured to a helmet liner which is considerably more complex than those of helmets issued later, in World War II. This relatively shallow helmet was, however, kept in service into World War II; not until 1944 did it begin to be replaced by the deeper Model 1944 helmet of the type still in service.

6 US Army tank crew helmet, World War II; common. This lightweight composition crash-helmet is heavily padded and pierced with ventilation holes. There is provision for earphones in the flexible leather earpieces. Note the snap-fastened strap that is shown here across the crown: this could be fastened across the straps of tank goggles to keep them in

place. A closely-fitting fabric "flying-helmet" having a lengthy skirt to protect the wearer's neck was issued for wear beneath the crash-helmet, but most tankers found this uncomfortable and preferred to supplement the crash-helmet with the knitted "jeep cap" or "beanie"; see (2) and (3) on *pages 92-93*.

7 Infantryman's steel helmet, Dutch Army, World War II. Very similar to the Italian helmet shown at (3), but with smaller metal studs, the Dutch helmet also closely resembled those worn by the Romanian and Soviet infantrymen. Dutch helmets, which are relatively common, are often found with oval bronzed metal badges showing the lion emblem of the House of Nassau.

8 Paratrooper's steel helmet, British Army, World War II; a rare type. This design was introduced c1942 as the result of examination and testing of German paratroop helmets. Earlier versions had rubber rims: the helmet shown is of the second type, with a turned rim incorporating a protective band of hard rubber, about 1in (25mm) wide. The leather neck and chin strap has a three-point fixing and incorporates a chin-cup to ensure that the helmet stayed firmly in place during jumping and combat. The shell of this helmet was used, with different liners, by despatch riders and AFV crews. It was succeeded by a rimless helmet—which is more common; the one shown is relatively scarce—and finally by a type with webbing neck and chin strap that was in service from 1944 until the present day.

As in other equipment, variants are found in helmet types. This US Army tankers' helmet differs from (6), above, in ventilation holes and details of ear-pieces.

9 Infantryman's steel helmet, Dutch Army, post-World War II; common. The major apparent variation from the earlier Dutch helmet shown at (7) is the rolled rim. Six rivets secure the liner and a simple leather chin-strap is fitted.

10 French Army tank crew steel helmet, World War II; common. Shallower than the infantryman's helmet (4) and lacking its comb, the AFV crews' helmet of the type introduced c1935 is readily recognisable by its external padded leather forehead protector. Not seen here are its broad leather chin straps, with accommodation for ear phones and buckling on the left.

1 "D" Type flying helmet of the Royal Air Force; issued for use in desert and tropic theatres during World War II. The pattern much resembled that of the "C" Type helmet shown at (4), but the "D" Type is made of khaki drill material and has a quilted neck flap for protection against sun and wind. It is shown here with a pair of Mk III flying goggles.

2 Type A-6 flying helmet of the United States Army Air Force, World War II. It is shown with the Type A-14 demand oxygen mask, which was designed to accept a modified microphone attachment.

3-4 Flying goggles of the Imperial Japanese Navy Air Force, World War II; a very rare item. The goggles are shown with their cleaning cloth (4), which is printed with instructions for care and maintenance of the goggles. Note that Imperial Japan had no Air Force as such; both the Imperial Army and Navy maintained their own air arms. The helmet shown with these items, although somewhat resembling the light drill flying helmets of Japanese aviators, is not an authentic piece and is intended only to display the goggles.

5 "C" Type flying helmet of the Royal Air Force, World War II; the standard issue brown leather helmet with chamois lining, introduced in the autumn of 1941, and the type most likely to be found by the collector in the United Kingdom. External wiring connects the earphones and the microphone to a universal jack plug. The goggles shown with this helmet are again the most common British type and were introduced in 1941. Both helmet and goggles of this type remained in service with the RAF into the 1950s.

6 "B" Type flying helmet of the Royal Air Force; the pattern worn from the mid-1930s and throughout the Battle of Britain of August-September 1940. Shown with the helmet is a pair of Mk IV goggles with polarised flip-down screens (shown in the "up" position) as a protection against the glare of bright sunlight. The microphone shown is from a later "D" Type oxygen mask. Note the external wiring that also serves to secure the goggles.

7 AN-H-15 flying helmet of the US Army Air Force, World War II; a light-weight helmet made of khaki

Flak helmet of the United States Army Air Force, World War II. Adapted from the GI steel helmet of 1942, this was issued to US bomber crews from 1943 onward. It was worn over the leather flying helmet, as shown, as a protection against small-calibre bullets and shell fragments. The hinged metal ear-flaps allowed room for headphones to be worn. To minimise the chance of icing at high altitudes, the steel helmet was sprayed with felt-based paint. Men of the US 8th Air Force flying from bases in the United Kingdom in the strategic air offensive against Germany were equipped with flak helmets and also body armour of "lobster tail" overlapping steel plates, produced by the famous Wilkinson Sword company of Great Britain.

drill, also called "chino", and designed for use in the Pacific Theater of Operations. A side view of this helmet, which was very popular with US pilots and aircrew, is shown here.

8 Air Ministry pattern Air Speed Indicator used by the Royal Air Force, c1939-1945 and later. This is an example of the standard instrument fitted to light aircraft and training aircraft such as the de Havilland Tiger Moth. Instruments like this—moderately easy to detach and conceal—were usually the first "souvenirs" to disappear from aircraft that had crashed.

9 Air Ministry pattern Climb and Descent Meter, used by the Royal Air Force on "blind flying" instrument panels.

10 Leather flying helmet of the type worn by men of the Royal Flying Corps, forerunner of the Royal Air Force, during World War I. This is a rare example, of top-quality material and still in excellent condition, of the early style of flying helmet, designed to give maximum protection from the elements when worn in the open-cockpit aircraft of the period. The style shown is that favoured by air gunners; pilots tended to favour a smaller, less all-enveloping helmet. Note the adjustment straps; not visible in this photograph are the toggles behind the ear flaps, which were used to secure the indispensable goggles.

11 Face mask issued to bomber crew members of the United States Army Air Force, World War II. These masks were issued to air gunners stationed in the open ports of the "waist", or similarly exposed positions, of the B-17 "Flying Fortress" and other heavy bombers. In service from 1942 until the end of the War, they were primarily intended to combat the danger of frostbite at high altitudes—the B-17 had a service ceiling of 35,000ft (10,670m)—and might be worn in conjunction with the F-2 electrically-heated flying suits and specially-insulated boots.

12 Mk IIIA goggles of the Royal Air Force; these were introduced in 1933 and remained in service until the introduction of the Mk IV goggles—shown on the "B" Type helmet at (6)—in 1940. These goggles are now an extremely rare item and are eagerly sought by collectors of aeronautica.

Cloth Insignia

Tony Walker

The collecting of cloth badges and insignia covers the entire field of 20th-century militaria, as well as much of the 19th Century, for then, too, embroidered badges were often used in place of metal for distinguishing rank, recognition of units and indication of long service. For the collector of 20th-century militaria, cloth insignia includes such items as shoulder patches, beret flashes, trade and qualification badges, rank insignia, shoulder titles, epaulettes, collar patches, naval cap tallies and armbands of all kinds.

Although space has forbidden their inclusion in this book (and the same consideration is likely to inhibit many collectors), flags and banners fall within this category also. Few British colours and banners come on to the market, since these are treasured regimental possessions. Nazi Germany generated an enormous number of flags and banners in 1933-45: these are eagerly sought by collectors, but although the smaller car pennants may still be found, they are, like much other Nazi militaria, extremely expensive. For the collector of Japanese material of this kind, the smaller "Rising Sun" battle-flags and the so-called "Kamikaze headbands" (*hachimaki*), bearing the red disc and a poetic inscription, are the most popular items but, again, rare and expensive.

Flags and banners apart, cloth insignia are mainly intended as badges of recognition or proficiency. Around the turn of the century, service dress began to replace full dress for general service wear. The British and American Armies adopted khaki (originated in India; the name comes from the word for "dust") which was worn during the Boer War (1899-1902) in South Africa by the British and in Cuba during the Spanish-American War (1898) by the Americans. Since this drab colour was worn for camouflage purposes, the wearing of polished metal insignia tended to destroy its effect. British units in South Africa cut the shoulder straps with embroidered county and regimental names from their scarlet or red tunics and sewed them to the puggarees of their helmets. The insignia of ranks below commissioned officer, usually white tape chevrons, were dulled and backed with khaki cloth. In the final stages of the Boer War, officers wore the same equipment as their men, so as not to provide an obvious target for Boer marksmen.

Thus, by the beginning of the 20th century, most armies had begun to utilise cloth insignia: the French on kepis and collar patches; the British even on officers' service dress, which bore regimental badges of dull bronze colour; and the Americans with arm of service badges.

The Brigade and Divisional system had been in use for a number of years by 1914, but when very large armies were put into the field, the universal use of service dress meant that an added identification needed to be placed on soldiers' uniforms. Signs to identify armies, corps and divisions, first developed by the British for use on transport, were worn on the upper sleeve—and in some cases on the back, so that a unit could be identified from the rear by the formation behind it in an advance. Because many line and territorial regiments had a great number of battalions, these were identified by a Roman numeral or a cloth flash worn on the upper left sleeve.

All armies by this time had their so-called "trade badges", the distinguishing marks of specialists. In the British Army, these were either brass or woven. The French Army's trade badges were either cut from cloth or embroidered. In the German Army of the period, distinguishing marks included both badges and the various coloured pipings (*Waffenfarbe* "arm of sevice colours"; a system preserved and expanded under the Third Reich) of shoulder straps, collar patches and headgear. The American forces of the period of World War I were distinguished also by coloured cords on the campaign hat. The first US unit to adopt a formation sign, in 1918, was the 81st Division, who wore a round patch showing the Carolina Wildcat on the upper left sleeve of the tunic. This was done on the unit's own initiative; but in October 1918 all units of the

Above: *US 8th Army Commander, General James A. Van Fleet—four-star rank insignia and 2nd Infantry Division patch—with British Field Marshal Viscount Alexander and staff on HMS* Ocean *off Inchon, Korea, June 1952.*

Far left: *US Marines and Japanese PoW on Iwo Jima, March 1945. Only an armband denotes the Military Policeman (MP).*

Left: *Paratroopers of US 101st Airborne Division—note the famous "Screaming Eagle" patch—prepare to board their aircraft for "D-Day", 6 June 1944.*

American Expeditionary Force were authorised to wear a badge or patch to distinguish the various divisions. It was at first painted on the helmet or, made from felt, sewn on the sleeve: there was no standard issue until after the end of World War I. (It is worth noting that the earliest use of cloth insignia was during the American Civil War (1861-65). General Phil Kearny of the Union Army, wishing to be able easily to identify his Corps' troops, ordered that each man should wear a red cloth patch on his cap. This became known as the "Kearny Patch". Other Corps adopted this idea, but few of the original patches survive.)

PATCHES OF US FORCES

American shoulder patches can generally be identified fairly easily, since they frequently carry the unit number. However, this is sometimes hard to decipher, because of the use of a rebus, or visual pun. For example, the 13th Army Corps wears a patch bearing a four-leafed clover, superimposed upon which is a triangle in red. The cloverleaf is in the form of a letter "X", the Roman numeral for 10, while the triangle represents the figure "3"; thus giving the equation 10 + 3 = 13. Other units have taken their insignia from heraldic beasts, as in the unicorn of the 13th Airborne Division; from heraldic puns, such as the 27th Division's design of the constellation Orion, a pun on the name of General O'Ryan, its commander in World War I; from plants or symbols of the area from which the unit was raised, as in the 24th Division's taro leaf, a plant common in Hawaii, its home base; or from pure heraldic symbolism, as in the patch of the European Theater of Operations, World War II, with its design of twin yellow and red lightning flashes breaking a chain, symbolising the Allied armies breaking the Nazi chains binding the nations of Occupied Europe.

Armored Divisions all wore a patch of the same design, consisting of a triangle divided into three segments: yellow at the top to denote cavalry; blue on the left to denote infantry; red on the right to denote artillery. Superimposed on this was a set of tank tracks bearing a cannon barrel, crossed with a red lightning flash. The whole represented the firepower of artillery, the attacking speed of cavalry and the fighting capability of the infantry. The divisional number was shown in the apex of the patch.

Most US Armored Divisions of World War II had nicknames, the best-known probably being the 2nd Armored Division's "Hell on Wheels", and tabs showing these nicknames were attached to the bottom of the patch.

The nicknames are: 1st Armored Division: "Old Ironsides", a reference to General Patton; 2nd: "Hell on Wheels"; 3rd: "Spearhead"; 4th: "Breakthrough", not authorised for wear; 5th: "Victory", unofficial; 8th: "Thundering Herd", "Iron Snake" and "Tornado", all unofficial tabs worn by World War II veterans; 9th: "Remagen", embroidered in green on a black rectangle and worn as a battle honour by veterans of Combat Command "B" of the 9th Armored, who captured the bridge over the Rhine at Remagen, March 1945; 10th; "Tiger", unofficial; 11th: "Thunderbolt", unofficial; 12th: "Hellcat", unofficial; 13th: "Black Cat", unofficial; 14th: "Liberator", unofficial.

At the present time (1983), only the 2nd and 3rd Armored Divisions are on active status. The 2nd Armored wear the patch on the left pocket of the fatigue uniform, rather than on the left shoulder, in deference to the wish of the unit's World War II commander General George S. Patton Jr, who wanted 2nd Armored's unit pride to be "close to the hearts of my men". It is worth noting also that the largest unit patch worn in the US Army is that of 1st Cavalry Division; (see (1) on *pages 126-127*).

During World War II, the US Army Air Forces wore, in addition to shoulder patches, squadron patches on their leather flying jackets. The patches were made of leather and the designs were very colourful, with winged skulls, red devils throwing lightning bolts, pugnacious-looking bulldogs with machine guns and other cartoon characters. The Walt Disney Studios designed a number of these patches, featuring such characters as Goofy, Dumbo the Flying Elephant and the Big Bad Wolf. All of these patches are now extremely rare and examples in mint condition bring very high prices.

Patches of the US Airborne Forces are an extremely popular subject, and numerous different versions are available at reasonable prices to collectors. Of the five Airborne Divisions of World War II, the 11th, 17th, 82nd and 101st saw combat, but the 13th Div. did not. The Parachute Infantry Regiments—all numbered in the 500 series; ie, the 501st, 503rd, 508th and so on—also wore patches, but these were unofficial and were usually worn on the pocket of the fatigue jacket. The rarer airborne patches of World War II, those of Airborne Command and Airborne Troop Carrier Command, are hard to find but, as with most cloth insignia, not prohibitively expensive.

In the 1960s the Vietnam War brought a multitude of US patches into being, the vast majority of them unauthorised types (see *pages 126-127*). Generally, a unit, having decided on a design for a patch, would order a quantity of between 20 and 100 from the nearest tailor shop to the base, and these would be hand-made on the spot. The quality of these patches varies: some are beautifully made and detailed; others, such as the "Sat Cong" ("Kill Cong") patches unofficially awarded to soldiers who had killed a Viet Cong or North Vietnamese soldier in combat are merely olive-drab discs with the wording crudely embroidered in the centre. US Special Forces in Vietnam had many different patches, which again vary in quality. Most designs showed a skull wearing a green beret, a dagger dripping blood, a striking cobra or a rattlesnake. These patches are now very rare and bring very high prices in the United States. Air Cavalry patches, bearing such designs as helicopters carrying cavalry sabres, winged skulls with the motto "Death From Above", or even the cartoon character Snoopy, are also difficult to find, especially for collectors in Europe.

The patches of the former Republic of South Vietnam are growing in popularity among collectors, since there were literally hundreds of different types, usually screen-printed onto cotton material or woven in silk. Provincial Reconnaissance Units (PRU) patches normally had a winged scimitar in the design: these units carried out infiltration and abduction missions in Viet Cong-held territory, with a view to bringing back prisoners for interrogation.

INSIGNIA OF THE THIRD REICH

The wealth of cloth insignia of the Third Reich (see *pages 134-135, 136-137*) has already been mentioned. The Nazi Party and all its ancillary organisations, as well as the fighting services, had distinctive uniforms and insignia which provide a fascinating field for the collector. There are many types of armband (*pages 138-139*), ranging from the well-finished examples worn by Nazi Party officials to the crudely-made bands lettered *Deutsches Volkssturm,* worn by the "home guard" defenders of Berlin in 1945.

There are also many variants of the Nazi Eagle cloth insignia, which was usually worn over the right breast pocket of the uniform (but on the upper left sleeve by the SS and Waffen-SS). These eagles were made in a number of different materials, including silk embroidery, usually known as "Bevo", gold and silver wire, and cotton. The background colours also vary, according to whether the uniform was a summer or winter type. The Luftwaffe design differed from the Army, Navy and SS types, inasmuch as the eagle's wings were in a natural flying position rather than being stylised. Reproductions of these are numerous: the collector should be cautious in purchasing.

Cloth insignia of the SS (*Schutzstaffel*) is extremely popular, but expensive. Collar patches with twin lightning-flash insignia were worn by most SS units, with the exception of SS Division *Totenkopf,* which wore an embroidered Death's Head. Other designs of collar patch were used by foreign volunteers in the SS, but these are rarely seen. Officers wore the insignia in embroidered silver wire; NCOs and ORs in light-grey cotton.

Cuff titles—see (9), *pages 134-135* —thin strips of black material worn round the lower left sleeve of the uniform, bore the name of the division or unit embroidered in silver wire or grey thread. These are very scarce, and they

are also being reproduced on a massive scale, to the same specification as the original in some cases, so caution is advised. Luftwaffe flying badges were made in cloth as well as metal and can be found occasionally, as can the sleeve badges worn by Radio Operators, Signallers, Flak Artillerymen and other specialists. So, too, can the collar patches worn by the German Army, usually on a background of *Waffenfarbe:* white for Infantry; pink for Panzer troops; red for Artillery and so on.

BRITISH INSIGNIA

British cloth insignia collecting, a rapidly expanding field, may be divided into three main categories: formation signs; shoulder titles; rank and tradesmen's insignia.

Formation signs (or shoulder flashes) are no longer worn, but during World War II there were many different types, some of which are illustrated on *pages 112-113, 114-115.* They first came into use during World War I and were introduced both as a means of identifying different units and to encourage *esprit de corps.* They went out of use after World War I, but with the outbreak of World War II they were reintroduced. The designs usually refer to the unit, although sometimes rather obscurely. Some refer to the area from which the unit was raised, such as the "HD" of the 51st Highland Division or the "TT" of the 50th (Northumbrian) Division, which stands for the rivers Tyne and Tees. Other units had heraldic devices: the 2nd Infantry Division, which was part of Northern Command, used the crossed keys from the coat of arms of the Archbishop of York. The flashes themselves were usually printed on cloth or embroidered on felt.

Cloth formation signs of British Airborne Forces are much sought after. The Parachute Regiment, formed in World War II, adopted Pegasus, the winged horse of Greek mythology, as its sleeve badge, which was embroidered in pale blue on a maroon felt background. Parachutist wings, worn on the upper right sleeve of the uniform consist of a white parachute in the centre of a pair of blue wings on a khaki background. The Special Air Service (SAS) wears a different-shaped wing, in shades of light and dark blue. The SAS beret flash is a dark blue shield on which is embroidered a winged dagger with a scroll, bearing the regimental motto "Who Dares Wins".

Shoulder Titles are perhaps the most popular area for new collectors, for they can still be found quite easily and there are a great number from which to choose. They consist of narrow strips of cloth, embroidered, or sometimes printed, with the name of the regiment or unit, and were sewn to the upper sleeves of the uniform just below the shoulder seam. Infantry regiments wore titles with scarlet backgrounds; Rifle regiments wore dark green titles embroidered in black. Royal Artillery units wore titles embroidered in red on a blue background.

There are many variations of British NCO and Warrant Officer rank insignia,

Above: *German Mountain Troops (Gebirgsjäger) — note embroidered "Edelweiss" proficiency arm badges — during World War II.*

Left: *Note the silver-thread cuff title of the* Germania *Regiment,* Waffen-SS; *these are now eagerly sought by collectors.*

Right: *British wounded return from Normandy, July 1944; the Leading Aircraftwoman wears the Red Cross armband common to all "medics".*

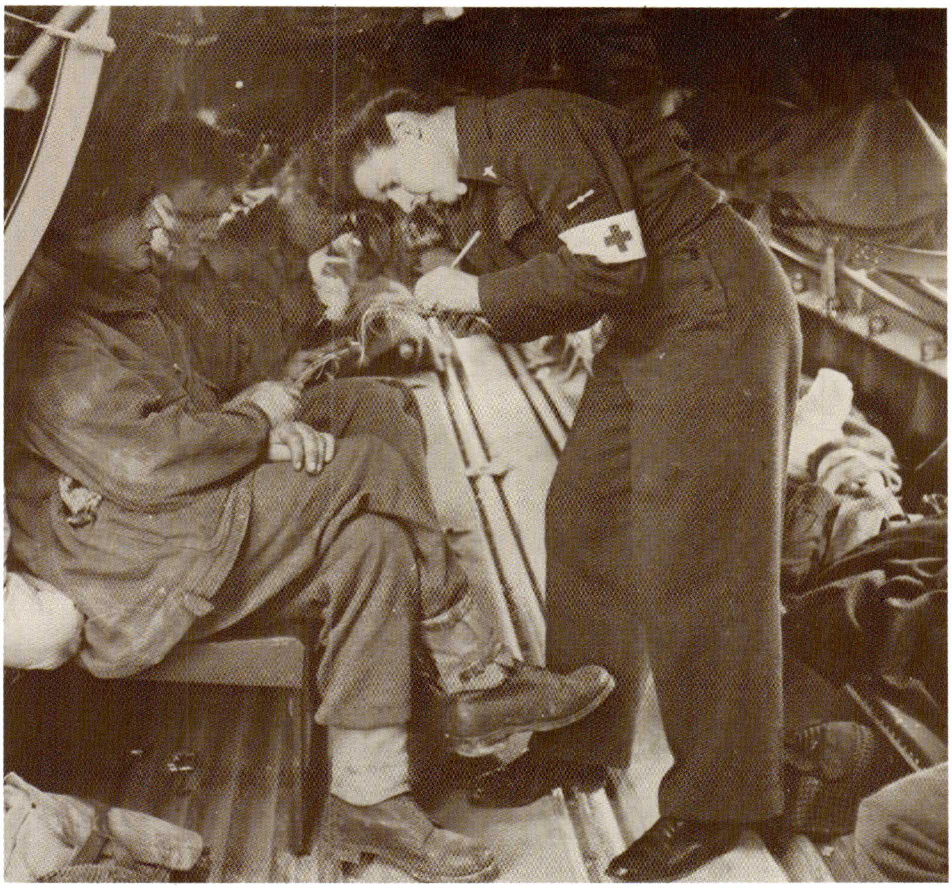

ranging from the crudely-printed chevrons seen during World War II, to the elaborately-embroidered sleeve badge of a Regimental Sergeant-Major, which bears the Royal Coat of Arms. Tradesmen's badges, worn on the lower right sleeve, include the insignia worn by such specialists as Farriers, Mechanics, Drivers, Snipers and, in the case of Scottish regiments, Pipers. The Royal Navy is rich territory for collectors (see *pages 116-117*), with very many types of cloth sleeve badges for the different ranks and branches of the service. One very popular field is the collecting of "cap tallies", the strips of black silk embroidered with the name of the wearer's ship in gold wire, worn around the cap.

A few collectors now specialise in the cloth insignia of other countries; notably Rhodesia (now Zimbabwe), whose unit patches of such elite forces as the Selous Scouts, Special Air Service and Greys Scouts are especially sought after.

FINDING AND KEEPING

To anyone contemplating a collection of cloth insignia, the field may at first seem overwhelming. However, once the prospective collector has decided which country's or service's patches are to be collected, the field narrows and the search can begin. The first rule of patch collecting is to gather as much information as possible. Read every reference book available on the subject (see *pages 206-207*) and join the clubs and societies catering for collectors of military cloth insignia. These publish regular magazines and newsletters, and often carry members' advertisements through which insignia can be bought, sold or exchanged.

Other sources of cloth insignia are militaria shops, antique fairs and flea markets, where it is possible to find bargains. Unfortunately, these are also the places where one is likely to find the reproduction items that may be bought by the unwary new collector. Reproduction Nazi militaria is being made now on an enormous scale; such items as SS cuff titles, collar patches and sleeve eagles are mass-produced in Austria to original specifications. British cloth items are also being made, although on a smaller scale, the most commonly found reproductions being insignia of Parachute and Commando units of World War II. American shoulder patches are being turned out by the hundred in Taiwan and Korea: only the rarer items are reproduced, such as the Vietnam-period Special Forces and Air Cavalry patches, the US Navy Underwater Demolition Teams and SEALS (Sea and Land Forces), and the rarer Airborne patches. Strange to say, the reproductions are often of better quality than the originals, for they are machine-woven rather than made by hand.

As well as reproductions of American patches, mention should be made of patches made up for units after they have returned from a combat posting. An example is the "Merrill's Marauders" patch. This was never worn: the unit's designation in World War II was 5307th Composite Unit (Provisional), and the patches were made in India after the Marauders' campaign was over. The "Son Tay Raider" patch, showing a PoW camp guard tower and wire fence with an arrow landing and taking off from the centre, is another example. The top of the patch bears a scroll with the words "Son Tay Raider", while another scroll at the bottom bears the legend "21 November 1970". This commemorates the abortive attempt by Special Forces troops to free American PoWs from the camp at Son Tay, Vietnam. The patches were privately made for distribution to the men taking part—and reproductions of these non-official patches have in turn come on to the collectors' market.

DISPLAY

Displaying cloth insignia is relatively easy. Smaller items, such as British shoulder titles and formation signs, may be mounted on a suitable cloth backing in old picture frames, which can then be hung on the wall. Take care however, not to hang them in direct sunlight, as this may cause the colours to fade. If space will not permit a wall display, albums may be purchased, the most suitable being binders with clear polythene inserts designed to hold postcard collections. Mount patches and insignia on white card, using "Blu-Tack" or some similar adhesive but *not* staples, for these will rust and stain the insignia. Slip the cards into the polythene inserts and the insignia will then be kept permanently clean and flat.

British Army Formation Signs of World War II and After

British Army formation signs, originally adopted during World War I to provide identification of a unit in a way that would not be undesirably explicit to unauthorised persons, are worn on the upper sleeve. The examples shown here date from World War II and later: the period covered has been specified wherever possible, but it should be noted that the wartime and post-war signs are often identical, or have only minor changes in design. Where no period is specified in the caption, it may generally be assumed that the sign dates from World War II. It should also be noted that just as signs differ in manufacture—printed, embroidered, or woven into felt, woollen or silk material—so may variations in colour and detail be found in signs relating to a single unit. Most signs have some symbolic or heraldic significance: subject to the limitations of space, an attempt has been made to explain the meaning of those that are not self-evident.

1 12th Infantry Brigade; a different sign—a fouled anchor in white, with yellow cable, on a blue diamond—was worn by HQ personnel. Such variations between the insignia worn by a formation's troops and its HQ personnel are not uncommon; see also (55).
2 4th Infantry Division, post-1945.
3 HQ Middle East Land Forces.
4 Southern Command (UK), Royal Signals, post-1945.
5 2nd Infantry Division; the crossed-keys badge, Archbishopric of York.

6 HQ Land Forces Hong Kong, post-1945; the Chinese dragon.
7 London District, Southern Command, post-1945.
8 49th (West Riding) Infantry Division; the polar bear crest was adopted during service in Iceland, 1941.
9 23rd Infantry Division; found also in other colours.
10 46th (North Midlands) Infantry Division.
11 GHQ Home Forces.
12 GHQ 2nd Echelon, 21st Army Group.
13 52nd (Lowland) Division; the unit was trained for mountain warfare

Cloth insignia collectors' interest in this SD jacket of a Captain, Royal Artillery, will centre on the embroidered Glider Pilot's wings on the left breast.

and wore a scroll with the word "Mountain" beneath the St Andrew's Cross shield.
14 South-Eastern Command (UK).
15 East Kent District, Eastern Command; design shows the "white cliffs of Dover".
16 3rd Infantry Brigade.
17 GHQ India.
18 9th Armoured Division; the powerful but pacific panda, it is said, symbolised the unit's role as a training formation only, 1941-44.
19 Sussex District, South-Eastern Command.
20 231st Independent Infantry Brigade; adopted during the defence of Malta, 1941-42.
21 GHQ Middle East Land Forces, post-1945.
22 6th Armoured Division.
23 8th Army Corps; adopted in 1943.
24 Gibraltar Garrison; the "key to the Mediterranean".
25 Wiltshire and Dorset District (pre-1944, Salisbury Plain District); the sign shows Stonehenge.
26 War Office Controlled Units; the crown differs on earlier signs of the post-1946 period.
27 Southern Command, Royal Army Service Corps; post-1945.
28 50th (Northumbrian) Infantry Division.
29 12th Signal Group.
30 39th (City of London) Signal Regiment.
31 Guards Armoured Division; the sign, adopted 1941, was that of the Guards Division of World War I.
32-33 1st and 4th Guards Brigade; post-war signs, with Roman numeral added to (31).
34 21st Army Group.
35 Formation sign, 14th Army.
36 Formation sign, 8th Army.
37 Cyprus District.
38 23rd Armoured Brigade; raised in Liverpool, the unit took the city's "Liver Bird" as its sign during World War II.
39 Gold Coast District.
40 Western Command (UK).
41 24th Independent Guards Brigade Group.
42 HQ Southern Command (UK); post-1945.
43 5th Infantry Division; "Y" signifies "Yorkshire".
44 Eastern Command (UK); 1944-47.
45 13th Army Corps.
46 HQ 21st Army Group.
47 Eastern Command, Royal Signals; 1944-47.
48 11th Army Corps.
49 42nd Armoured Division; disbanded 1943. The sign is wrongly shown: it was worn as a diamond.
50 Malaya Command, post-war; the Malayan *kris* (dagger).
51 5th Army Corps; the Viking ship was chosen to mark the unit's service in Norway, 1940.
52 British Troops in France; based on the sign of 21st Army Group—see also (12) and (46)—which formed the basis of the British Army of the Rhine (BAOR) in 1945.
53 Allied Land Forces, South-East Asia; worn only by British forces.
54 Anti-Aircraft Command, post-1945.
55 Scottish Command, post-1945; this was the sign worn by troops: HQ staff wore the same sign but with a central horizontal black band.
56 43rd (Wessex) Infantry Division (embroidered pattern); see also *pages 114-115*, (49).

British Army Formation Signs of World War II and After

British Army formation signs: see introductory paragraph to captions on *pages 112-113.*

1 East Anglian District, post-1946; similar insignia, in gold on blue, worn by East Anglian Training Brigade Group, post-1947.
2 6th Infantry Brigade, post-World War II; the emblem of key crossed with bayonet was also worn by 5th Infantry Brigade.
3 1st Army Corps; formed September 1939 – the spearhead symbolises its position as the "first part" of the British Expeditionary Force sent to France at the beginning of World War II.
4 17th Indian Division; original sign, 1943, was on a khaki background. The yellow of the sign shown was retained post-war when the sign was adopted by the 17th British Division in Malaya.
5 3rd Infantry Division.
6 Airborne Division.
7 34th Anti-Aircraft Brigade, Royal Artillery, post-1945.
8 5th Anti-Aircraft Division.
9 HQ Northern Ireland District, post-1948; earlier signs show a bird on a nest or (later) a gate, both in white on green.
10 ANZAC (Australia and New Zealand Army Corps) Brigade, Singapore.
11 10th Army Corps.
12 47th (London) Infantry Division; the sign is a rebus (visual pun) on "Bow Bells", since popular tradition has it that only those born within hearing distance of the bells of the church of St Mary-le-Bow, Cheapside, City of London, can claim to be true "Cockneys", or natives of London.
13 47th (2nd London) Division, Royal Signals.
14 Southern Command, Infantry, post-World War II; the various branches of service within Southern Command wore this sign, based on the stars of the Southern Cross constellation, in their own colours: variations are shown on *pages 112-113*, (4), (27) and (42).
15 1st Canadian Army, formed 1942; of this Army's two Corps, 1st Corps wore a similar diamond, all scarlet, and 2nd Corps an all-blue diamond.
16 13th Infantry Division, from 1945; the sign, dating from World War I, bears the "lucky" horse-shoe to counter the "unlucky number".
17 Territorial Army, Troops (printed pattern); see also (35).
18 6th Anti-Aircraft Division.
19 79th Armoured Division.
20 51st (Highland) Infantry Division.
21 255th Signal Squadron, Bahrain; post-World War II.
22 54th (East Anglian) Infantry Division, post-World War II.
23 Singapore District, post-1945.
24 55th (West Lancashire) Infantry Division; note the five petals, sepals and stamens of the Rose of Lancaster, and the five leaves on each stem.
25 HQ 1st Army Group Royal Artillery, post-1945.
26 Air Formation Signals, post-World War II; the colour bands symbolise sky, sea and land and the unit's role in all three elements.
27 Southern Command, Royal

27

28

29

30

31

32

33

34

35
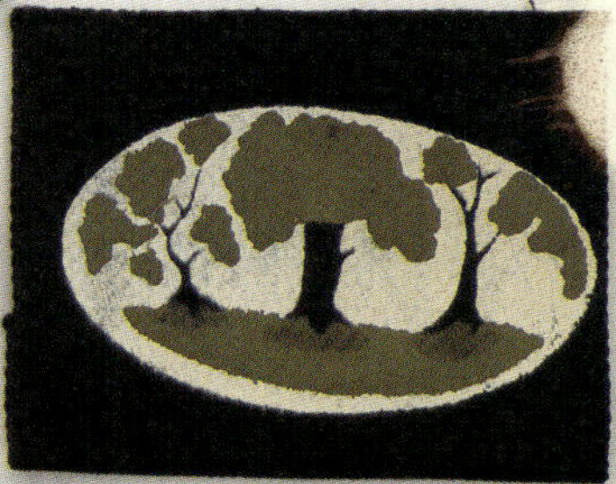
36
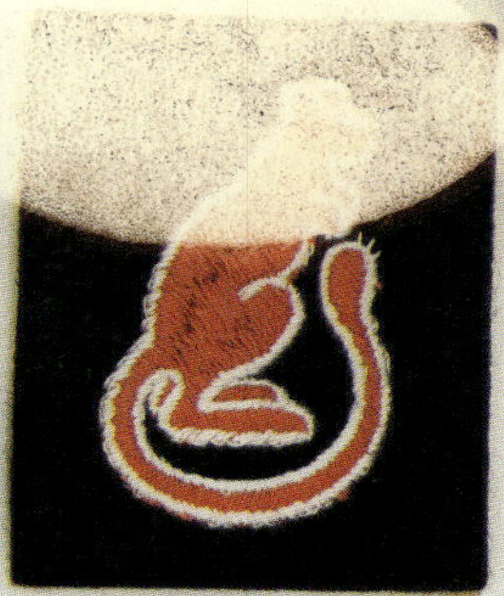
37

38

39

40

41

42

43

44

45

46

47

48

49

50

51

52

Officer's leather jerkin, British Auxiliary Territorial Service (ATS; women's service); the lining bears cloth insignia of Allied (and German) units, c1944-45.

Electrical and Mechanical Engineers (REME), dating from post-1945; see also (14).
28 8th Army (printed pattern).
29 4th Indian Division.
30 56th (London) Infantry Division; the sign shows "Dick Whittington's cat", the mythical beast that helped its master become Lord Mayor of London.
31 Northern Command, from 1947; the earlier badge showed a green apple on a blue diamond.
32 (Above) 63rd Gurkha Infantry Brigade; (below) 99th Gurkha Infantry Brigade. The sign with crossed kukris is common to all Gurkha formations.
33 15th Signal Regiment.
34 Glider Pilot; specialist's arm badge.
35 Territorial Army, Troops (woven pattern); see also (17).
36 12th Army Corps; the sign shows the "oak, ash and thorn" used by Rudyard Kipling to symbolise the Sussex region where the unit was raised in World War II.
37 7th Armoured Division; the jerboa of the famous "Desert Rats".
38 30th Army Corps.
39 9th Army Corps (second pattern); earlier in World War II the Corps wore a sign bearing a "nine-lived" black cat in a posture of defiance.
40 Royal Artillery; unfortunately this patch is shown inverted.
41 63rd Signal Regiment (Aldershot District).
42 The War Office, from 1946.
43 53rd (Welsh) Infantry Division.
44 Signals Training Regiment, post-World War II.
45 HQ Combined Operations; the World War II sign with eagle, anchor and "tommy gun" symbolises the special forces' wartime operations by air, sea and land.
46 Midland West District; post-1945.
47 78th Infantry Division.
48 11th Armoured Division.
49 43rd (Wessex) Infantry Division (woven pattern); see also *pages 112-113*, (56).
50 1st Infantry Division; since this is the first division of 1st Army Corps, which readopted its "spearhead" sign – see (3) – when reformed in the 1950s, the divisional sign represents the point of the spear.
51 1st Canadian Army Group Royal Artillery; the 1st Army sign, see (15), with red zig-zag added.
52 North Midland District (Northern Command), post-1945; the sign shows the legendary Robin Hood of Sherwood Forest, N. Midlands.

1-2 Badges authorised in 1979 and current, worn on right sleeve of No 2 uniform (red) and No 1 uniform (gold) by Leading Weapons Engineering Mechanic (Radio). Stars denote proficiency.
3 No 2 uniform badge (as 1-2) of Weapons Engineer Mechanic (Ordnance). Letter "O", below, signifies "Ordnance"; star, above, denotes able rate.
4 Jacket lapel badge, authorised 1965 and now obsolete, worn on No 2 uniform by Underwater Weapons Chief Petty Officer 1st Class; crown and star show rank.
5 Badge, 1960-79, worn on right sleeve of No 1 uniform by Seaman Gunner (Control), Royal Naval Reserve. "MS" denotes qualified in minesweeping.
6 Jacket lapel badge, 1960-79, worn on No 2 uniform by Chief Petty Officer, Royal Naval Reserve.

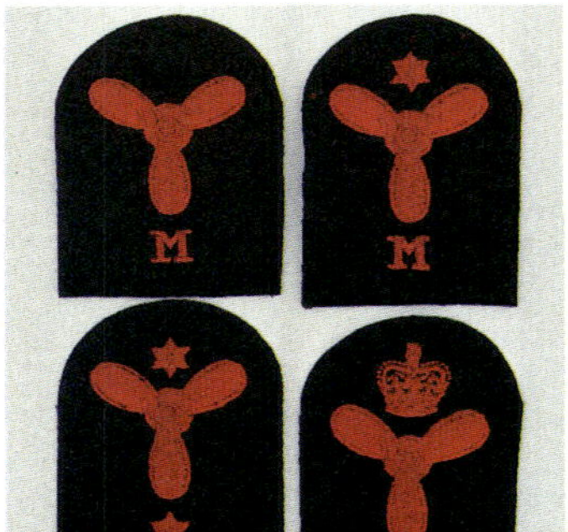

Sleeve badges, 1979 current, of Marine Engineering Mechanics (Mechanical), RN. Left to right: 2nd Class; 1st Class; Leading Mechanic; Petty Officer.

7 Badge, 1949-56, worn on right sleeve of No 1 uniform by Petty Officer Airman Fitter (Airframes).
8 Gold wire badge, 1940-47, worn on right sleeve of No 1 uniform by Seaman-Wireman. "L" denotes "Electrical".
9-10 Gold wire badges, 1955-79, worn on right sleeve of No 1 uniform by Electrical Mechanic (Air) 2nd Class (9) and Leading Electrical Mechanic (Air).
11 Gold wire badge, authorised 1965 and current, worn on right sleeve of No 1 uniform by Medical Technician (Radiographer), 1st, 2nd, 3rd Class.
12 Title worn on right sleeve, above any branch badge, by all Chief Petty Officers, POs and Ratings on H.M. Royal Yacht *Britannia*.
13 Title worn on No 2 uniform by Master-at-Arms on board *Britannia*.
14 Title worn at top of sleeve on navy blue battledress jacket; normally worn by beach parties.
15 Sleeve title currently worn by Commando-trained RN personnel attached to Royal Marine units.
16 Sleeve title currently worn by all Royal Naval Reserve ratings.
17 Badge, authorised 1870 and obsolete, formerly worn on left sleeve by (i) Petty Officer 2nd Class, a rank abolished in 1913; (ii) Boy Petty Officer at naval training establishment.
18 Badge, 1951-79, worn on lapel of No 2 uniform by Chief Petty Officer, Surveying Recorder 1st Class.
19 Badge, 1960-69, worn on right sleeve of No 2 uniform by Electrical

Mechanician (Air) 5th Class.

20 Badge, 1948-76 (but "king's crown" dates this example to 1948-53), worn on right sleeve of No 2 uniform by Petty Officer Mechanician 2nd/3rd Class.

21 Gold wire badge, 1976, current, worn on No 1 uniform by basic rate Mine Warfare Rating.

22 Badge, 1976 current, worn on right sleeve of No 2 uniform by Petty Officer, Mine Warfare.

23 Gold wire badge, 1948 current, worn on right sleeve of No 1 uniform by Petty Officer, Writer.

24 Gold wire badge, 1980 current, worn on right cuff of No 1 uniform by men of Subsunk Parachute Assistance Group.

25-26 Badges, 1959 current, worn on right cuff of tropical uniform (25) and No 2 uniform (26) by RN personnel who have passed RAF Para School.

27 Gold wire badge, 1890-1948 and now fairly rare, worn on right sleeve of No 1 uniform by Armourer's Mate.

28 Pilot's wings, 1925 current, worn on left forearm above rank stripes, by Royal Navy Officer Pilots.

29-30 Pilot's wings, *c.* 1940-48, worn 1·5in (38mm) above left breast pocket on white uniform (29) and blue serge working dress (30) by Royal Navy Rating Pilots. Only about 150 Rating Pilots were awarded wings.

31 Gold wire badge, 1967-77, worn on right sleeve of No 1 uniform by Ordnance Electrical Mechanician (Air) Acting 4th or 5th Class.

32 Gold wire badge, 1975 current, worn on right cuff of No 1 uniform by Airborne Missile Aimer (Fleet Air Arm helicopter aircrew).

33 Gold and silver wire badge, 1970 current, worn 3in (76mm) above cuff of Nos 1 and 2 uniforms by Fleet Chief Petty Officer.

34 Gold wire badge, 1909-58, worn on right sleeve of No 1 uniform by Leading Signalman.

35-37 Badges, 1942 current, worn by Rating Aircrew: (35) gold wire, on cuff of left sleeve of No 1 uniform; (36) left breast pocket of tropical tunic; (37) cuff of left sleeve of No 2 uniform.

38 Badge, 1969-79, worn on lapels of No 2 uniform by Chief Control Electrician.

39 Cap Badge, 1970, current, of Fleet Chief Petty Officer.

40 Gold and silver wire cap badge, 1953 current, RN officer.

41 Brass cap badge, World War II period, Chief Petty Officer.

42 Brass cap badge, World War II period, Petty Officer.

43 Gold wire badge, 1951 current, worn on lapels of No 1 uniform by Chief Petty Officer, Cook.

44 Gold wire badge, 1951-64, worn on lapels of No 1 uniform by Stores Chief Petty Officer (Victualling).

45 Slip-on rank insignia, worn on naval action shirt and "woolly pully" by Lieutenant, Royal Navy Reserve.

46 Gold wire badge, 1947-51, worn on lapels of No 1 uniform by Chief Petty Officer, Stoker Mechanic.

47 Gold wire badge, 1934 current, worn on lapels of No 1 uniform by Chief Radio Supervisor.

48 Rank badge, 1913 current, worn on left sleeve of No 2 uniform by Petty Officer.

US Army Divisional Patches of World War II

1 1st Inf Div, "Big Red One"; original patch, World War I, was reputedly made from a piece of red cloth from a captured German cap.
2 2nd Inf Div, "Indianhead"; the design resulted from a competition, for a prize of 40 francs, in World War I.
3 3rd Inf Div, "Marnemen"; design recalls the division's three great battles in World War I, when its tenacity in defence won it the title "Rock of the Marne".
4 4th Inf Div, "Ivy Division"; a rebus design: the Roman numerals for four are "IV" and the patch shows four ivy leaves.
5 6th Inf Div.
6 23rd Inf Div, "Americal"; nickname a contraction of "America" and "New Caledonia", where they saw action in World War II.
7 7th Inf Div, "Hourglass"; design represents two crossed "7s".
8 8th Inf Div, "Pathfinders".
9 9th Inf Div; the patch is called the "Salerno Butter Cookie", from the shape and the unit's first major battle, 1943.
10 10th Mountain Div; "Mountain" tab was worn above unit patch.
11 29th Inf Div, "The Blue and the Gray"; colours of design, a Korean good-luck sign, signify the unit was raised from states opposed in the Civil War.
12 24th Inf Div; originally a Hawaiian National Guard unit, hence taro leaf design.
13 27th Inf Div, "New York"; design combines monogram "NYD" and stars of the Orion constellation, a pun on "O'Ryan", name of commander in 1917-18.
14 34th Inf Div, "Red Bull"; shaped like an *olla*, Mexican water container, with a red bull to mark training in New Mexico.
15 18th Airborne Corps; an overseas cap patch.
16 35th Inf Div, "Santa Fé"; design based on markers used on the old Santa Fé Trail.
17 30th Inf Div, "Old Hickory"; the nickname was that of President Andrew Jackson, in whose home state, Tennessee, the unit was raised.
18 25th Inf Div, "Tropic Lightning".
19 26th Inf Div, "Yankee Division".
20 32nd Inf Div, "Red Arrow"; design symbolises claim that unit pierces every enemy line it faces.
21 33rd Inf Div, "Prairie"; raised in Illinois, unit took design from symbol used in World War I to mark poison-gas containers.
22 43rd Inf Div, "Red Wing".
23 A rare patch: the first worn by US Rangers, 1942, before they were formed into separate battalions.
24 36th Inf Div, "Texas"; design melds Indian arrowhead and "T".
25 31st Inf Div, "Dixie Division"; with "double D" design.
26 40th Inf Div, "Sunshine".
27 38th Inf Div, "Cyclone".
28 91st Inf Div, "Lone Pine".
29 44th Inf Div; "Double 4" design.
30 45th Inf Div, "Thunderbird"; an Amerind symbol reflecting the high proportion of American Indians in unit raised mainly from Oklahoma National Guard.
31 37th Inf Div, "Buckeye"; design from the state flag of Ohio.
32 Tab worn above unit patch by personnel qualified as Rangers.

33 Shoulder flash, 3rd Ranger Battalion.
34 5th Inf Div, "Red Diamond".
35 42nd Inf Div, "Rainbow".
36 41st Inf Div, "Sunset".
37 63rd Inf Div, "Blood and Fire".
38 65th Inf Div, "Battleaxe".
39 66th Inf Div, "Black Panthers".
40 69th Inf Div; design, a stylised "6" and "9", recognisable by TV fans as the patch worn in the "Sergeant Bilko" series!
41 70th Inf Div, "Trailblazers".
42 71st Inf Div.
43 75th Inf Div.
44 76th Inf Div.
45 77th Inf Div, "Liberty"; the unit of New York State.
46 78th Inf Div, "Lightning".
47 79th Inf Div, "Lorraine"; nickname and design, Cross of Lorraine, stem from service in France, during World War I.
48 80th Inf Div, "Blue Ridge".
49 81st Inf Div, "Wildcat"; first patch authorised, in World War I.
50 83rd Inf Div; monogram "OHIO", where the unit was recruited.
51 84th Inf Div, "Railsplitters".
52 85th Inf Div, "Custer Division"; originally trained at Fort Custer.
53 86th Inf Div, "Blackhawk".
54 87th Inf Div, "Acorn Division".
55 89th Inf Div, "Rolling W" or "Mid-West"; "W" inverted becomes "M".
56 90th Inf Div, "Tough 'Ombres"; "T" and "O" for Texas and Oklahoma, where unit was raised. Nickname means "Tough Guys".
57 96th Inf Div, "Deadeye".
58 98th Inf Div, "Iroquois"; head of an Iroquois Indian, original inhabitants of New York State where unit was raised, on colours of the House of Orange, commemorating Dutch settlers of New Amsterdam, later New York.
59 92nd Inf Div, "Buffalo"; formed in World War I, one of two divisions of black soldiers, called "buffaloes" by American Indians.
60 97th Inf Div, "Trident".
61 93rd Inf Div; second of the black divisions, see (59). The design features a steel helmet of French type (*Casque Adrian*), commemorating the division's distinguished service in France in World War I.
62 94th Inf Div.
63 100th Inf Div, "Century Division".
64 95th Inf Div, "Victory"; "V" stands for both "Victory" and Roman numeral for "five".
65 103rd Inf Div, "Cactus Division".
66 28th Inf Div, "Keystone"; unit was raised from Pennsylvania National Guard and patch is in the shape of the keystone on the seal of Pennsylvania State.
67 104th Inf Div, "Timberwolf".
68 99th Inf Div.
69 A very rare red "Airborne" tab, worn above the patch of 84th Inf Div when it was briefly redesignated an airborne unit in the 1950s.
70 106th Inf Div, "Golden Lion"; design from insignia of college fraternity of original commander.
71 102nd Inf Div, "Ozark"; another rebus: "Z" within "O" with geometrical "arc" below.
72 Most infantry divisions that were redesignated airborne units from 1942 onward adopted new shoulder patches; 82nd and 101st Divs retained their former designs, but wore these tabs above them.

Cloth Insignia of US Forces, World War II

1 Flash worn by accredited press with US forces, World War II.
2 Constabulary in Europe, worn by members of Occupation Forces in Germany on Military Police duty, immediately after World War II.
3 442nd Regimental Combat Team, formed of Japanese-Americans.
4 Allied Forces HQ, North Africa, worn by US and British personnel.
5 Official US War Photographer (see 1).
6 *Stars and Stripes,* official daily newspaper of US Army, was printed in all theatres of war.
7 US Navy Amphibious Forces, worn on upper left sleeve.
8 North African Theater of Operations; shaped like dome of mosque.
9 HQ, European Theater of Operations; worn by troops attached to Gen Eisenhower's HQ in UK and France.
10 European Command (formerly Supreme Headquarters, Allied Expeditionary Force, SHAEF); originally on a black background.
11 Handmade variation on (1); all press patches are now very rare.
12 US Army Amphibious Forces (see 7).
13 Women's Auxiliary Army Driver Corps, worn before formation of WAAC. This very rare patch is shown with the "Hollywood" tab; tabs are found for other areas.
14-15 Occupational Therapy Specialist and (15) Occupational Therapy Apprentice, US Red Cross.
16 1st Marine Amphibious Corps Aviation Engineers.
17 Amphibious Engineer Brigade, worn by 2nd and 409th Engineer Special Brigades.
18 US Military Mission to Moscow; a very rare patch, worn by US Army staff of the US-Russian Aid Program.
19 HQ, Strategic Air Command; currently worn by crew of flying Command Post—a converted Boeing 707 continually airborne in case SAC's HQ is knocked out.
20 Antilles Department, US Army; design shows Morro Castle, Puerto Rico.
21 3rd Armored Div, "Spearhead".
22 Flash worn by civilians working in US Army Service Clubs, World War II.

Royal Netherlands Army, World War II: ribbon of Oorlogsherinnerings Kruis (War Commemorative Cross) with bronze stars indicating two bars; cloth and metal titles.

23 General HQ, Southwest Pacific, worn by US personnel at General MacArthur's HQ, World War II.
24 The Armor School, worn by trainees at US Army Tank Training Center, Fort Knox.
25 39th Inf Div, "Delta Division"; recruited in Mississippi delta states.
26 7th Service Command, responsible for induction and training camps in north-central states, 1941-45.
27 General Headquarters Reserve, World War II.
28 9th Service Command, as (26), covering Far West states.
29 98th Inf Div; "subdued" version of flash on *pages 118-119*, (8).
30 Ports of Embarkation, worn by men of Transportation Corps.
31 9th Army, World War II.
32 Army Transportation Corps, a breast pocket patch.
33 US Air Force Squadron Patch; not identified and unofficial. Presumably stationed in Germany—eagle wears Tyrolean garments.
34 Patch worn by US Forces, Dominican Republic, in multinational force assisting Dominican government against communist-led revolt, 1965. "FIP" and "OEA" are Spanish acronyms for "International Peace Force" and "Organisations of American States".
35 14th Army; a "ghost" unit which was never actually formed.
36 7th Army Corps, World War II.
37 8th Army, World War II.
38 5th Army Corps, World War II.
39 1st Army Corps, World War II.
40 European Theater of Operations, Advanced Base; World War II.
41 Hawaiian Department; US Army.
42 War Correspondent; an unofficial variation on official patch (1).
43 US Army Forces Pacific Ocean Area; design shows Southern Cross with arrow pointing NW—to Japan.
44 2nd Security Zone California; a very rare patch dating from "invasion fever" in California immediately after attack on Pearl Harbor, December 1941.
45 US Army Service of Supply, also known as Army Service Force.
46 War Correspondent; official patch to be worn on shoulder loop.
47 US Army Replacement and School Command; shown reversed: blue segment should be on left.
48 Western Pacific Command, US Army.
49 Victory Task Force; patch with letter and morse-code "V" worn by troops taking part in displays across the USA in 1942 to raise funds for Army Emergency Relief Program.
50 1st Army; symbolised by first letter of alphabet.
51 5th Service Command (see 26), for Kentucky, Ohio, West Virginia and Indiana.
52 London Base Command; a rare patch—never officially authorised, although several variations exist—worn by personnel of LBC before it became European Theater of Operations (see 9).
53 A further variation on the War Correspondents' flash.
54 Armored Force; patch worn by instructors and pupils at US Army Tank Training School. This was unit designation until Armored Forces were reorganised into separate Armored Divisions.

US Airborne Forces Insignia, World War II to Vietnam

1 Airborne Command; worn in World War II by paratroopers not assigned to Airborne divisions and by those assigned to headquarters and training duty.

2 1st Allied Airborne Army; formed 1944 and including British 1st Airborne Div, US 82nd and 101st Airborne Divs and Free Polish Parachute Brigade. A British-made patch, with design embroidered on felt; US-made patch is embroidered throughout.

3 9th Troop Carrier Command, US Army Air Force.

4 9th "Ghost" Airborne Div; one of the non-existent units of US First Army Group (FUSAG), a decoy organisation formed in the US in 1944 to mislead the Germans concerning Allied invasion plans. Details of "Ghost" units are still classified. Although the patches were officially approved, they were probably not worn.

5 18th "Ghost" Airborne Div, FUSAG.

6 21st "Ghost" Airborne Div, FUSAG.

7 135th "Ghost" Airborne Div, FUSAG.

8 11th Airborne Div, "Angels"; nick-name earned in drop on Japanese PoW camp at Los Banos, Luzon.

9 17th Airborne Div.

10 13th Airborne Div.

11 17th Airborne Div; gold wire embroidered patch of the type sometimes made up for officers.

12 101st Airborne Div, "Screaming Eagles"; design remembers "Old Abe", mascot of Union Army's "Iron Brigade" in the Civil War.

13 11th Air Assault Div; very rare patch of the forerunner of modern helicopter-borne units.

14 18th Airborne Corps.

15 Forty-eight star American Flag sleeve patch, worn by US soldiers for ready identification when entering enemy-occupied areas. There were several different types, including cloth and oilcloth armbands.

16 82nd Airborne Div, "All American", has seen more action than any other US Airborne unit and is presently a Rapid Deployment Force (RDF).

17 187th Airborne Regimental Combat Team (RCT); see also (19).

18 307th Airborne Engineer Battalion; unofficial patch worn on pocket of a fatigue uniform.

19 Variation on 187th Airborne RCT patch (17), in United Nations colours of pale blue and white.

20 541st Parachute Infantry Regiment; patch of post-war design for wear on blazer or casual jacket.

21 511th Parachute Infantry Regiment; word "Angels" refers to unit's part in Los Banos drop with 11th Airborne Div (see 8).

22 508th Airborne RCT.

23 Special Forces, better known as "Green Berets", deep-penetration and counter-insurgency teams first formed in late 1950s. Patch shaped like Indian arrowhead, with sword and three lightning bolts denoting capacity to strike from land, sea or air.

24 Beret flash worn by 5th Special Forces Group Vietnam, Studies and Operations Group.

25 Beret flash of HQ staff, 82nd Airborne Div.

26 Cap patch, Airborne Forces.

27 Patch worn by men of 101st

Airborne Div who have passed RECONDO (RECONaissance CommanDO) course.

28 8th Ranger Battalion, shoulder flash.

29 9th Ranger Battalion.

30 1st Battalion, 75th Infantry; a direct descendant of 5307th Composite Unit (Provisional), the famous "Merrill's Marauders".

31 101st Airborne Div, compare with (12): subdued patch made for Vietnam service, where it was found that bright patches made aiming points for snipers.

32 82nd Airborne Command and Control, worn by HQ staff and emblematic of their many functions.

33 F Troop, 4th Cavalry, Hunter Killer Team; unofficial patch, Vietnam.

34 82nd Airborne Div, "Devil Brigade", unofficial pocket patch.

35 Airborne Tab, worn over unit patch to denote parachute qualification.

36 Special Forces School (formerly the John F. Kennedy Center).

37 509th Airborne Infantry Regiment, worn by men of 509th attached to 8th Inf Div.

38 Special Forces pocket patch.

39 Special Forces, subdued patch (see 23, 31).

40 Ranger Tab, worn as (35).

41 L Company, 75th Infantry, Vietnam.

42 5th Special Forces Group (Vietnam) Command and Control; pocket patch, like most Special Forces patches, unofficial and not authorised for wear outside combat zone.

43 Special Forces background oval, worn as backing to parachute wings over left breast pocket.

44 Beret flash, 1st Special Forces.

45 Military Assistance Command Vietnam, Studies and Observation Group.

46 503rd Parachute RCT.

47 501st Parachute Infantry Regiment pocket patch; motto is famous "paratrooper yell".

48 Company A, 3rd Battalion, 5th Infantry Div, pocket patch.

49 508th Parachute Infantry Regiment, "Red Devils"; part of 82nd Airborne Div, World War II.

Tunic (pattern from 1964) of a Major, Armoured Reconnaissance, Bundeswehr (West German Army). Paratroop insignia with gold oak-leaves for more than 50 jumps; Close Combat/Ranger badge; gold Bundeswehr-Leistungsabzeichen, awarded for general excellence in work, marksmanship and sport.

1 1st Air Force, United States Army Air Force (USAAF), World War II. This shoulder sleeve patch, like all similar USAAF patches of World War II, is in blue and gold. 1st Air Force served in the American Theater of War, mainly employed on training duties.

2 2nd Air Force; also employed only in the American Theater of War during World War II.

3 3rd Air Force; employed in anti-submarine duties off the US coasts in World War II.

4 7th Air Force; employed in the Pacific Theater of War, with campaign participation in the Western Pacific, Ryukyu Islands and China Offensive.

5 8th Air Force, the "Mighty Eighth"; most famous of all USAAF forces, 8th Air Force was based in the UK and took part in the strategic air offensive against Germany, World War II.

6 12th Air Force; employed in the North African and Italian Theaters, World War II.

7 Shoulder sleeve patch worn by personnel at USAAF Headquarters, Washington D.C., World War II.

8 20th Air Force; served in the Pacific Theater, where it flew the B-29 Superfortresses that spearheaded the aerial assault on Japan. It was from the B-29 "Enola Gay" (named after his mother), that Colonel Paul Tibbets, 393rd Bombardment Squadron, 20th Air Force, delivered the first atomic bomb on Hiroshima, 6 August 1945.

9 Weather Specialist, USAAF; this badge, authorised in January 1943, was worn 4in (100mm) above the cuff of the right sleeve on all uniforms, or on the left breast pocket of fatigue dress, by airmen on meteorological duties.

10 Photographic Specialist, USAAF; as (9), worn by photographers, photographic interpretation specialists and camera repairmen.

11 Cushion cover bearing the emblem of the US Air Forces. These items were sold in PX (Post Exchange) stores on US bases all over the world, and were popular souvenirs for the serviceman to send to the folks at home. Many different designs may be found, including covers showing tanks, jeeps, battleships and various kinds of aircraft. They are attractive collectables, not rare and usually found at reasonable prices.

12 6th Air Force; the galleon in the design was the insignia of US Caribbean Defense Command, to which 6th Air Force was attached on anti-submarine duties during World War II.

13 11th "Ghost" Div, US Army; the significance of such patches is explained on pages 122-123 (9). Design symbolises the face of a clock, with the "eleventh hour" segment in black.

14 1st Army Corps; served in the Pacific Theater, World War II.

15 14th "Ghost" Div.

16 22nd "Ghost" Div.

17 46th "Ghost" Div.

18 2nd Army Corps; the combination of the American Eagle and British Lion commemorates the Corps' service on the Western Front in World War I.

19 50th "Ghost" Div.

20 55th "Ghost" Div.
21 141st "Ghost" Div.
22 119th "Ghost" Div.
23 13th Air Force, USAAF; served in the Pacific Theater, World War II.
24 130th "Ghost" Div.
25 59th "Ghost" Div.

US Army Officer's winter-pattern tunic, World War II. Brass badges on lapels: (above) US national insignia; (below) service branch badge of Engineers. Cloth insignia: eight bars on right sleeve denote four years' overseas service; shoulder patch (part visible), 5th Army; medal ribbons: Bronze Star; unidentified; American Campaign; European-African-Middle Eastern Campaign; Victory, World War II; Army of Occupation; Korean Service; United Nations, Korea.

26 157th "Ghost" Div.
27 High School Victory Corps; the military appearance of this flash is misleading: it was the sleeve insignia worn by a World-War-II organisation of American high-school pupils who collected re-usable scrap for the war effort.
28 Combat Infantry Badge; authorised on 27 October 1943 for wear above the left breast pocket by officers and men of the US Army who had distinguished themselves in ground combat, this badge, in silver and blue enamel was also made in cloth—seen here—for wear on field uniform.
29 20th Army Corps.
30 108th "Ghost" Div.
31 4th Army; served only as a training unit in the USA during World War II.
32 Women's Auxiliary Army Corps tab. The WAAC was formed on 16 May 1942 to train women for tasks that would enable men to be freed for the fighting forces. In 1943 it was incorporated in the regular Army as the Women's Army Corps (WAC). Women who had enlisted before the incorporation wore WAAC tab on left sleeve to show length of service.
33 7th Army; spearheaded the invasion of Southern France, 1944.
34 US Navy Construction Battalions (CB, hence "Seabees") flash. Formed in 1942 to perform building and maintenance duties in combat zones, there were some 340 of these units by August 1945.
35 US Air Force pocket tab, for wear on flying suits.
36 US Air Force pocket tab, for wear on fatigue dress.

Cloth Insignia of US Forces in the Vietnam War

During the Vietnam War, US units serving in that country from 1961 to 1972 initiated the production of shoulder and pocket patches on a vast scale. These patches were not officially authorised, although in many cases they had the approval of commanding officers. The small Special Forces ("Green Berets") units originated a large number of patches; common motifs being a skull wearing a green beret—see *pages 123-123,* (45)—or a dagger dripping blood. Also popular are the patches of Air Cavalry units. Collectors of Vietnam-made unofficial patches should bear in mind that, because of the lack of reference material on these insignia, reproduction insignia have been produced on a large scale in Taiwan and Pakistan. There are also a number of "fantasy" insignia, for units that never existed, in production. One such piece (45) is shown here.

1 1st Cavalry Div; official sleeve insignia. After serving as infantry in World War II, the division was reformed with the 11th Air Assault Div as an Airmobile unit, equipped with helicopters.
2 Airmobile patch; cloth version of the Airmobile Badge (3), for wear on a flying suit.
3 Airmobile Badge: metal breast badge worn by men who have completed the Airmobile course.
4 1st Cavalry Div; "subdued" patch, worn in combat zone, where a brightly-coloured patch might provide an aiming point.
5 Military Assistance Command, Vietnam (MAC-V); a "subdued" version of (10).
6 Yankee Air Pirate; unofficial patch, quoting communist propaganda, worn by US Air Force aircrew on missions over North Vietnam.
7 US Army Vietnam "subdued" version of (13).
8 1st Aviation Brigade; a "subdued" patch made in Vietnam.
9 20th Engineer Brigade.
10 Military Assistance Command, Vietnam; official patch.
11 Regimental Combat Team 2, 1st Cavalry Div; a Vietnam-made patch in which only the colour of the divisional patch (1) varies.
12 1st Field Force, US Army Vietnam.
13 US Army Vietnam.
14 269th Combat Aviation Battalion, "Black Barons"; patch for Air Cavalry flying suit.
15 US Marine Corps Recon; worn by Airborne Marines, Vietnam.
16 Ranger-Jungle Fighter; worn by Philippine contingent of Free World Force, Vietnam.
17 1st Recon Battalion, USMC; an unofficial patch worn at bases.
18 Tracker; patch worn by members of US Army's Dog Platoons, or "K-9 units", which were used to track Viet Cong infiltrators.
19 Military Assistance Advisory Group, Vietnam; worn early in the Vietnam War, when US personnel acted only as "advisers".
20 Scout Ranger; official tab.
21 Scout Ranger; patch worn by Filipino troops serving with the US Army in Vietnam.
22 199th Infantry Brigade.
23 169th Infantry Brigade.
24 21st Infantry Regiment; pocket patch.

24 25 26 27 28 29 31 33 34 35 37 30 32 36 38 39 40 41 42 43 44 45 46 47

25 101st Airborne Div; an unofficial patch.
26 Senior Parachutist; cloth version of metal breast badge.
27 Special Forces; unofficial.
28 3rd Armored Cavalry Regiment.
29 1st Bn, 327th Infantry; beret flash. The 82nd and 101st Airborne Divisions now wear berets (maroon and blue respectively) and each battalion has its own beret flash.
30 2nd Bn, 327th Infantry; beret.
31 75th Infantry; beret flash and metal badge.
32 3rd Bn, 187th Infantry; beret.
33 1st Bn, 501st Infantry; beret.
34 2nd Brigade, 101st Airborne Div.
35 1st Special Forces; beret flash and Special Forces badge. The badge is worn on the shoulder loops by officers, who wear rank insignia on beret flash.
36 2nd Squadron, 17th Cavalry.
37 1st Bn, 502nd Infantry; beret.
38 2nd Bn, 502nd Infantry; beret.
39-40 Shoulder insignia of Companies Kand N, 75th Infantry, Vietnam, 1969-1970.
41 Special Forces; silver-thread embroidered pocket patch. On it are (top left) 11th Air Assault "Sky Soldiers" badge, unofficial; (centre) 5th Special Forces Group (Vietnam) paratroop wing, worn unofficially by members of Studies and Observation Group (and now made in the USA and sold as "Mercenary Jump Wings"); (top right) Master Para Wing, Army of South Vietnam—also worn on right breast pocket by parachute-qualified US personnel.
42-43 Shoulder insignia of Companies I and J, 75th Infantry—as (39-40).
44 40th Armored Bn; pocket patch.
45 Reproduction pocket patch, apparently inspired by movie *Apocalypse Now*, in which motto appeared on helicopters.
46 South Korean Ranger; qualification badge worn by Republic of Korea volunteers in Vietnam.
47 82nd Airborne Div; "subdued".

Current West German insignia. Left to right, top to bottom: shoulder straps, Brigadier-General and Colonel; cap badges, Army officer and Air Force officer; shoulder-straps, Vice-Admiral and Lieutenant-Commander; rank insignia, Chief Petty Officer; Air Force specialist's badges, Cruise Missile Technician and Signals Technician; rank insignia, Lieutenant-Colonel.

1-28 Cloth insignia of the French Foreign Legion (Légion Etrangère).

1-5 Examples of the type of *écusson* ("shield"), a lozenge-shaped cloth badge, worn on the upper left sleeve by men of the Legion from the late 1940s to the present day. Green on navy blue/black is a distinguishing colour scheme of the Legion's insignia, although the Legion's official colours are red and green. (3), (4) and (5) bear regimental cyphers of the type common during the 1960s.

6 Non-Commissioned Officer (NCO) type écusson with the grenade badge of the Légion Etrangère embroidered in yellow silk; an unissued item, awaiting the addition of rank chevrons as shown in (11) and subsequent examples.

7 Officer's écusson of the type common in the early 1950s, the date of this hand-embroidered example from Indochina (Vietnam).

8 Rank chevrons of Sergent Chef (Chief Sergeant; Sergeant-Major), Légion Etrangère.

9 Rank chevrons and écusson of Caporal (Corporal), 3è Compagnie du 4è Régiment Etranger d'Infanterie (3rd Company, 4th Regiment of Legion Infantry), dating from 1930s.

10 Legion Officer's standard écusson of recent make: compare the quality of the gold-wire grenade with that shown at (11).

11 Rank chevrons and écussons of Caporal; officer's pattern and thus privately purchased. Note that rank chevrons are worn on both sleeves; the écusson only on the left. Compare with (12).

Service tunic, 1960s-1970s, of a Major, 3è REI (3rd Regiment of Legion Infantry); the rank is denoted by four gold-wire bands on shoulder boards. Note regimental patches on lapels, and green tie, the latter peculiar to the Legion. Metal insignia on right breast: Qualified Parachutist's wings; US Presidential Unit Citation; regimental insignia on green leather fob. Three cordes fourragère (lanyards) on left shoulder mark unit awards to 3è REI: Légion d'Honneur; Medaille Militaire; Croix de Guerre. Medal ribbons on left breast: Légion d'Honneur; Croix de Guerre 1939-45; Croix du Combattant; Colonia Medal; World War II Commemorative Medal; Moroccan Order of Ouissam Alaouite.

12 Caporal; standard pattern.
13 Caporal Chef (Chief Corporal); gold chevron above rank chevrons denotes the senior rank.
14 Caporal Chef of Cavalry; note colour change from gold to silver with change of arm.
15 Caporal Chef; note three re-enlistment chevrons beneath écusson.
16-17 Shoulder titles, Légion Etrangère: NCO (16) and Officer (17), both in flat wire thread.
18 Rank chevrons and Specialist's écusson of Legion Medical Corporal, worn on right sleeve of uniform.
19 Rank chevron and Specialist's écusson of Pioneer Légionnaire 1st Class.
20-22 Rank chevrons and écusson of (20) Sergeant; (21) Senior Sergeant-Major; (22) Sergeant; the latter is hand-made insignia dating from 1950s; all Legion Infantry.
23-24 Midnight blue shoulder boards with rank insignia and Legion badge, worn on parade dress by all ranks below Caporal Chef from 1948 to the present.
25-26 Shoulder boards for ranks Caporal Chef to Sergent Chef; (25) with Radio-Telegraphist's qualification insignia.
27 Lieutenant's shoulder board: mess dress pattern.
28 Major's shoulder board.
29-36 Collar patches worn on uniform of British pattern, with British rank insignia, by NCOs of the Polish Air Force in Britain, 1939-45. Red and white edging on (29) and (32) signifies an officer cadet in training; silver and white edging on (31), (35) and (36) signifies the wearer has passed officer training. Ranks shown are: (29) Flight Sergeant; (30/31) Sergeant; (32/33/34) Lance Sergeant; (35) Corporal; (36) Aircraftsman.
37-56 US Army sleeve insignia of World War I.
37 Private First Class (PFC), Ordnance Corps.
38 Gunner Second Class, Coastal Artillery Corps.
39 PFC, Signal Corps.
40 PFC, Artillery.
41 PFC, Engineer Corps.
42 Quartermaster Sergeant, Quartermaster Corps.
43 Coxswain, Coastal Artillery Corps.
44 Enlisted Man without specialist's rating, Squadron 344, Aviation Section, Signal Corps.
45 Master Engineer Senior Grade, Tank Corps; compare with earlier type of Tank Corps insignia shown at (54).
46 PFC, Air Corps.
47 PFC, Cavalry Corps.
48 Corporal, Ordnance Corps.
49 PFC, Quartermaster Corps.
50 PFC, Medical Corps.
51 Sergeant First Class, Motor Transport.
52 Master Electrician, Signal Corps.
53 Assistant Band Leader.
54 PFC, Tank Corps; note that this is a patch of an earlier type than (45) and shows a head-on view of a stylised tank.
55 Telephone Communications: the provenance of the patch is not known; it may have been improvised in the field.
56 Engineer, Coastal Artillery Corps.

Imperial German Shoulder Boards and Straps, 1900-1918

1 Service dress shoulder strap (*Feldachselstück*) of Medical General (*Generalarzt*), Prussia, 21 September 1915 on.
2 Parade uniform epaulette of Medical Officer ranking as Captain (*Stabsarzt*), Prussian, c1914.
3 Service dress shoulder strap of *Stabsarzt*—see (2)—Marine Infantry, c1900.
4 Specialist's badge, on field-grey (*feldgrau*) of World War I, of Medical Sergeant-Major (*Sanitätsfeldwebel*).
5 Service dress shoulder strap of Captain (*Hauptmann*), Foot-Artillery Regiment No 13 (Hohenzollern) (*Fussartillerie Regiment Nr 13 (Hohenzollernsches)*), 1888 on.
6 Service dress shoulder strap of Captain (*Hauptmann*), Regional Infantry Reserve, Reserve Inspectorate of Karlsruhe (*Provinzial-Landwehr-Infanterie der Landwehr-Inspektion Karlsruhe*), 29 June 1912 on.
7 Service dress shoulder strap of an Other Rank (OR), Engineer Section (*Pionier-Detachement*) of Far Eastern Colonial Troops (*Ostasiatische Kolonialtruppen*), c1900.
8 Service dress shoulder strap of Lieutenant-Colonel (*Oberstleutnant*), Emperor Alexander's Grenadier Guard Regiment No 1 (*Garde-Grenadier-Regiment Kaiser Alexander Nr 1*), 12 July 1888 on.
9 Service dress shoulder strap of *Stabsarzt*—see (2)—of Prussian Army, 21 September 1915 on.
10 Service dress shoulder strap of Second Lieutenant (*Leutnant*), Infantry Regiment No 18, 1915 on.
11 Parade uniform epaulette of Army Corps Veterinary Officer (*Korpsstabsveterinär*) ranking as Major.
12 Parade uniform epaulette of Paymaster (*Zahlmeister*) ranking as Second Lieutenant, Baden.
13 Hussar's shoulder cord (*Husarenschnur*) of brown material.
14 Parade uniform shoulder strap (*Schulterklappe*) of OR, Emperor Francis's Grenadier Guard Regiment No 2 (*Garde-Grenadier-Regiment Kaiser Franz Nr 2*), 1866.
15 Parade uniform shoulder strap of Veterinary Officer (*Unterveterinär*) ranking as Senior NCO, Prussia, 11 June 1912 on.
16 Parade uniform epaulette of Second Lieutenant (*Leutnant*), Alsatian Engineer Battalion No 16 (*Elsässisches Pionier-Bataillon Nr 16*), 1871 on.
17 Service dress shoulder strap of Second Lieutenant, Brandenburg Infantry Regiment No 92, 1871-95.
18 Service dress shoulder strap of OR, Infantry Regiment No 18, from 18 March 1910.
19 Parade uniform epaulette of Second Lieutenant, Regional Infantry Reserve, Reserve Inspectorate of Berlin, 1912 on.
20 Parade uniform shoulder strap of OR, 2nd Baden Grenadier Regiment Emperor William I No 110, with button of regiment's 7th Company; 1869 to 1915.
21 Parade uniform shoulder strap of Under-Officer (*Offizierstellvertreter*), Royal Bavarian Lifeguards.
22 Parade uniform shoulder strap of OR (one-year volunteer), Emperor

Alexander's Grenadier Guard Regiment No 1; 1856-1915.

23 Parade uniform shoulder strap of OR, Infantry Regiment Margrave Lewis William (3rd Baden) No 111 (*Infanterie-Regiment Markgraf Ludwig Wilhelm (3 Badisches) Nr 111*), 1902 to 1915.

24 Service dress shoulder strap of First Lieutenant (*Oberleutnant*), Count Schwerin's Infantry Regiment No 14 (3rd Pomeranian) (*Infanterie-Regiments Graf Schwerin Nr 14 (3 Pommersches)*).

25 Service dress shoulder strap of Second Lieutenant Infantry Regiment No 111 (3rd Baden)—later as (23)-1871-88.

26 Service dress shoulder strap of Paymaster (*Zahlmeister*) ranking as Second Lieutenant, Württemberg, 1888 on.

(Above) General officers' shoulder strap and epaulette with mourning bands; (below) General officers' parade uniform epaulette and service dress shoulder strap.

27 Service dress shoulder strap of Military Construction Official (*Regierungsbaumeister*) ranking as Major, Prussia, 1888 on.

28 Parade uniform epaulette of Baden Police-Inspector (*Polizei-Komissar*).

29 Parade uniform shoulder strap of OR, Field Artillery Detachment, Far Eastern Colonial Forces, c1900.

30 Service dress shoulder strap of Captain (*Hauptmann*), Supply Regiment No 14 (Baden), 21 September 1915 on.

31 Service dress shoulder strap of Director of Music (*Obermusikmeister*) ranking as First Lieutenant, Infantry Regiment von Lützow No 25 (1st Rhenish) (*Infanterie-Regiment von Lützow Nr 25 (1 Rheinisches)*), 1912 on.

32 Service dress shoulder strap of Captain, Foot-Artillery Regiment No 14 (Baden), 21 September 1915 on.

33 Parade uniform shoulder strap of Under-Officer (*Offizierstellvertreter*), Charles Antony von Hohenzollern's Fusilier Regiment No 40 (*Füsilier Regiment Karl Anton von Hohenzollern Nr 40*), March 1910 on.

34 Service dress shoulder strap of Second Lieutenant, Grand-Duke's Field Artillery Regiment No 14 (1st Baden), 1907 to September 1915.

35 Service dress shoulder strap of Captain, Grand-Duke of Mecklenburg's Rifle Battalion, 1888 on.

36 Hussar's shoulder cord (*Husarenschnur*) of silver wire; gold button.

37 Parade uniform shoulder strap of OR attached to 12th Army Corps (*XII Armeekorps*).

Imperial German Shoulder Boards and Straps, 1900-1918

1 Shoulder strap (*Schulterklappe*) worn on service dress by an Other Rank (OR) of the Foot-Artillery Inspection Board, (*Fuss-Artillerie-Prüfungs-Kommission*), from 21 September 1915 onward.

2 Epaulette worn on parade uniform by a Major (*Major*) of the Prussian General Staff, up to World War I.

3 Cloth specialist's badge, on the field-grey (*feldgrau*) cloth of the German service dress of World War I, worn on the right upper arm, (or on the left upper arm if some other badge took precedence on the right) by members of the Signal Troops attached to each Infantry Company of the German Army during World War I.

4 Epaulette worn on parade uniform by Colonel (*Oberst*) Freiherr von Lüttwitz, Prussian General Staff, before World War I.

5 Epaulette worn on parade uniform by a General (*Generaloberst*) acting in the rank of Field Marshal (*Generalfeldmarschall*), from 1911 onward.

6-7 Shoulder strap worn on service dress (*Feldachselstück*), and (7) epaulette worn on parade uniform, by a General (*Generaloberst*) acting in the rank of Field Marshal (*Generalfeldmarschall*), of the Royal Bavarian 8th Infantry Regiment of Grand-Duke Frederick of Baden (*Königlich Bayrisches 8 Infanterie-Regiment Grossherzog Friedrich von Baden*), up to 1911. Note the light-blue thread intertwined with the silver cord: these are the Bavarian national colours.

8 Shoulder strap worn on service dress by an aide-de-camp (*Flügeladjutanten*) to a General of the Prussian Army, from 12 July 1888 onward.

9 Shoulder strap worn on service dress by a Colonel (*Oberst*) of the Infantry Regiment Margrave Lewis William (3rd Badenian) No 111 (*Infanterie-Regiment Markgraf Ludwig Wilhelm (3 Badisches) Nr 111*), from 21 September 1915 onward.

10 Shoulder strap worn on service dress by a Bandmaster (*Musikmeister*)—a specialist with officer's rank equal to that of Lieutenant—of the 2nd Baden Grenadier Regiment of Emperor William I No 110 (*2 Badisches Grenadier-Regiment Kaiser Wilhelm I Nr 110*), from 18 May 1912 onward.

11 Shoulder strap worn on service

Officers' service dress shoulder straps (Feldachselstück) *of a Baden infantry regiment, with (right) field-grey covering adopted on the Western Front in 1914-15.*

dress by First Lieutenant (*Oberleutnant*) of the Infantry Regiment Margrave Lewis William (3rd Badenian) No 111 (*Infanterie-Regiment Markgraf Ludwig Wilhelm (3 Badisches) Nr 111*), from 24 May 1902 to 21 September 1915.

12 Shoulder strap worn on service dress by a Veterinary Lieutenant (*Oberveterinär*) of the Prussian Army, from 12 July 1888 onward.

13 Specialist's arm badge, metal on field-grey cloth, of a German Army Sharpshooter-Machine Gunner, World War I.

14 Shoulder strap worn on service dress by a Brigadier-General (*Generalmajor*) of the Prussian Army, from 12 July 1888 onward.

15 Specialist's arm badge, worn on the pre-1915 service dress tunic, of a Medical Sergeant-Major (*Sanitätsfeldwebel*).

16 Shoulder strap worn on field-grey greatcoat by an OR of the 2nd Baden Grenadier Regiment of Emperor William I No 110 (*2 Badisches Grenadier-Regiment Kaiser Wilhelm I Nr 110*), from 1910 onward.

17 Shoulder strap worn on service dress by a Second Lieutenant (*Leutnant*) of the Grand Duke's Field Artillery Regiment (1st Baden) No 14 (*Feldartillerie-Regiment Grossherzog (1 Badisches) Nr 14*), from 7 October 1907 onward.

18 Shoulder strap worn on service dress by a First Lieutenant (*Oberleutnant*) of the East Asian Expeditionary Force (*Ostasiatischen Expeditionscorps*). This was the force dispatched to China in 1900, under Field Marshal Count Alfred von Waldersee, to join the multi-national expedition to relieve the foreign legations at Pekin, which were besieged by the Chinese insurgents known in the West as "Boxers".

19 Epaulette worn on parade uniform by a Major (*Major*) of the 3rd Baden Field Artillery Regiment No 50 (*3 Badisches Feldartillerie-Regiment Nr 50*), from 1888 onward.

20 Shoulder strap worn on service dress by Grand Duke Frederick II of Baden as a General (*Generaloberst*) acting in the rank of Field Marshal (*Generalfeldmarschall*) of the 1st Baden Life Dragoon Regiment No 20 (*1 Badisches-Leib-Dragoner-Regiment Nr 20*), from 1911 onward.

21 Epaulette worn on parade uniform by a Second Lieutenant (*Leutnant*) of an unidentified unit.

22 Epaulette worn on parade uniform by a First Lieutenant (*Oberleutnant*) of the Westphalian Supply Battalion No 7 (*Westfälisches Train-Bataillon Nr 7*), from 1866.

23 Shoulder strap worn on service dress by a Warrant Officer (*Deckoffizier*), Marine Engineer (Torpedo Branch), Imperial German Navy.

24 Shoulder strap worn on service dress by King George of Saxony as a Field Marshal (*Generalfeldmarschall*) in the Old Württemberg Infantry Regiment (3rd Württemberg) No 121 (*Infanterie-Regiment Alt-Württemberg (3 Württembergisches) Nr 121*), before the outbreak of World War I.

1-3 Unit insignia of non-German volunteers serving with *Waffen-SS:* (1) provenance uncertain, probably Finnish troops; (2) *21st Waffen-Gebirgs-Division der SS "Skanderbeg" (Albanisches Nr 1)*, Albanian troops; (3) *20th Waffen Grenadier-Division der SS (Estnisches Nr 1)*, Estonian troops.
4 Rank insignia of Second Lieutenant (*Leutnant*); type introduced in 1942, colour printed on cotton, for wear on camouflage and special combat clothing.
5 Badge of *Kyffhäuserbund*, association of World War I veterans.
6 Insignia as (1-3), *Freiwilligen-Legion SS Norwegen*, Norwegian.
7 As (4); Colonel (*Oberst*).
8 Unit insignia, Polish Army.
9 Title worn around left cuff by Army personnel serving at Adolf Hitler's HQ (*Führerhauptquartier*). A rare item, hand-stitched in silver thread on black cotton, of the type worn from January 1941.

*Jacket of special field-grey combat uniform for crews of tank-destroyers and self-propelled guns in armoured (*Panzer*) units. Shoulder straps, with rose-pink* Waffenfarbe *of armoured units, show rank of Sergeant (*Feldwebel*). Death's-head (*Totenkopf*) collar patches show unit's traditional link with Death's Head Hussars (*Leibhusaren-Regiment 1 und 2*) and must not be confused with similar insignia of* Waffen-SS. *Medal ribbons: Iron Cross 2nd Class; Russian Campaign, 1941-42. Badges: Infantry Assault award; Wound Badge in black metal.*

10 National emblem (*Hoheitszeichen*) as worn on right breast of tunic by officers of armoured units.
11 As (10); Navy (*Kriegsmarine*).
12 Army Standard Bearer's arm shield, worn on right upper arm. Main colour in shield is "arm of service colour" (*Waffenfarbe*); in this case white, for Infantry.
13 *Hoheitszeichen* of type worn from 1944 on khaki uniform by men of Air Force (*Luftwaffe*).
14 Rank insignia of Senior Private (*Oberschütze*), left upper arm.
15 Flying Personnel badge; worn on left forearm by aircrew NCOs and men of Luftwaffe not qualified for Pilot, Observer or Radio Operator badges.
16-17 *Hoheitszeichen*, Army; (16) printed type, (17), embroidered type for officers.

18 *Hoheitszeichen,* Afrika Korps.
19 Felt embroidered badge worn on right upper arm by Mountain Troops (*Gebirgsjäger*); also found in machine-woven silk.
20 Felt embroidered badge worn on right upper arm by men of rifle formations trained as ski-troops (*Ski-Jäger*).
21-42 Trade and Specialist sleeve badges of Army and Luftwaffe:
21 Teletype Operator, Luftwaffe.
22 Qualified Radioman, Luftwaffe.
23 Qualified Telephonist, Luftwaffe.
24 Ordnance Sergeant-Major (*Feuerwerker*), Army.
25 Medical Orderly, Luftwaffe.
26 Medical Orderly (*Sanitätsunterpersonal*), Army.
27 Driver, Luftwaffe.
28-30 Signallers (*Nachrichtenpersonal*), other than men of Signals Corps, Army; lightning *blitz* in *Waffenfarbe* of Cavalry (28), Rifles (29), Infantry (30).
31 Signals Equipment Administrator, Luftwaffe.
32 Air Raid Warning Personnel, Luftwaffe.
33 NCO, Radio Direction, Luftwaffe.
34 Radio Operator (*Funkmeister*), Army; note that badge worn at angle of 90° to that shown here.
35 Ordnance NCO (*Waffenfeldwebel*), Army.
36 Qualified Farrier (*Geprüftes-Hufbeschlagpersonal*), Army; note badge worn in inverted position to that shown here.
37 Mechanised Transport Equipment Administrator, Luftwaffe.
38-40 As (28-30); *Waffenfarbe* of Mechanised Supply (38), Artillery (39), Armour (40).
41-42 Qualified Helmsman, Engineer Assault Boats (*Steuermann*), Army: silver embroidered anchor (41); metal badge on cloth (42).
43-50 Metal-on-cloth badges of Petty Officers (*Maate*), Navy; worn on left upper arm of parade jacket or short overcoat: (43) Driver; (44) Administrative Clerk; (45) Bandsman; (46) Gunner; (47) Signalman; (48) Radio Telegraphist; (49) Teleprinter Operator; (50) Gunnery Engineer.
51 Rank insignia worn on upper left arm by Army Corporal (*Obergefreiter*) with more than six years' service; silver embroidered stripes and metal star.
52 As (51) but without star: Corporal with less than six years' service.
53 As (52); Lance-Corporal (*Gefreiter*).
54-55 Specialists' arm badges of Transport NCO (*Schirrmeister*) of Luftwaffe (54) and Army (55).
56 Specialist's badge of Gun Layer, Artillery (*Richtkanonier*), Army.
57-58 Cap insignia of a Luftwaffe officer: (57) Cockade (*Reichskokarde*) with oakleaf wreath and wings, worn on cap band; (58) Eagle, worn above cockade. Note that the swastika that should depend from the eagle's claws is missing. ORs wore a similar cockade but without black cloth backing, and metal eagle.
59-62 Luftwaffe rank insignia, worn on both upper sleeves of flying uniform: (59) Sergeant (*Unterfeldwebel*); (60) Second Lieutenant (*Leutnant*); (61) Warrant Officer (*Oberfeldwebel*); (62) rank insignia of Major (*Major*).

Shoulder Straps, Lanyards and Belts of the Third Reich

1 Army Marksmanship Lanyard for NCOs and men: aluminium cord with alloy shield; loop above shield buttoned below right shoulder strap on parade, reporting, guard and walking-out uniforms. Example shown is basic grade, with shield of 1936-39 pattern.
2 As (1); 12th Grade, the highest; aluminium cord with gilt bindings; gilt shield of post-1939 pattern; three gilt acorns.
3 Air Force (*Luftwaffe*) Marksmanship Lanyard: grey-blue silk with interwoven silver thread and alloy shield; worn as (1). Basic grade shown: like the Army, *Luftwaffe* award had twelve grades.
4 Shoulder strap (*Schulterklappe*) worn on tunic and greatcoat by Infantry Private (*Schütze*), 14th Infantry Regt; white piping is "arm of service colour" (*Waffenfarbe*) of Infantry. Coloured piping (and later backing) for identification was in use in the Prussian Army by the mid-19th Century; *Waffenfarbe* were codified in the Imperial Army's Dress Regulations of 15 September 1915 and reached the height of complexity under the Third Reich.
5 Shoulder strap of Company Sergeant-Major (*Feldwebel*), metal pip and number.
6 Metal gorget (*Ringkragen*) with phosphorescent buttons, eagle and lettering, worn around neck by Military Field Police (*Feldgendarmerie*) on duty. This gave them the pejorative nickname of "Chained Dogs" (*Kettenhunde*).
7 Shoulder strap, Second Lieutenant (*Leutnant*), Aircrew or Parachutist (gold-yellow *Waffenfarbe*), *Luftwaffe*.
8 Shoulder strap, First Lieutenant (*Oberleutnant*), 2nd Cavalry Regt; gold-yellow *Waffenfarbe*.
9 Shoulder strap, Captain (*Hauptmann*), 9th Cavalry Regt.
10 Shoulder strap as (7), rank of Major.
11 Shoulder strap, Lieutenant-Colonel (*Oberstleutnant*), 13th Cavalry Regt.
12 Shoulder strap, Lance-Sergeant (*Unteroffizier*), Army Cavalry School (*Heeres-Kavallerie-Schule*).
13 Shoulder strap, Sergeant (*Unterfeldwebel*), Luftwaffe War School.
14 Shoulder strap, Captain (*Hauptmann*), 31st Artillery Regt; bright-red *Waffenfarbe*.
15-17 Shoulder straps of (15) Major (*Major*), (16) Lieutenant-Colonel (*Oberstleutnant*), (17) Colonel (*Oberst*); Artillery (*Heeres-Artillerie-Wesen*).
18 Collar patch (*Kragenspiegel*), Sergeant (*Unterfeldwebel*), Luftwaffe.
19 Collar patch, Aircraftsman (*Flieger*), *Luftwaffe*.
20 Collar patch, Army Administration Official (*Wehrmachtsbeamter*) of Technical Service, officer's rank.
21 Collar patch, Engineer officer.
22 Collar patch as (20), Cavalry.
23 Collar patch, Cavalry officer.
24 Collar patch as (20), Armour.
25 Collar patch, Infantry officer.
26 Shoulder strap, Lieutenant-Colonel (*Oberstleutnant*), 100th Motorised Infantry Regt; grass-green *Waffenfarbe*.
27-28 Shoulder straps of (27) Major (*Major*) and (28) Second Lieutenant (*Leutnant*); Motorised Infantry.

29-30 Shoulder straps of (29) Captain (*Hauptmann*) and (30) Second Lieutenant (*Leutnant*); Artillery (*Heeres-Artillerie-Wesen*).
31 Shoulder strap, Captain (*Hauptmann*), Army Tank School; rose-pink *Waffenfarbe*.
32 Shoulder strap, Major (*Major*), Armour.
33 Shoulder strap, Major (*Major*), 43rd Signal Regt; lemon-yellow *Waffenfarbe*.
34 Shoulder strap, Second Lieutenant (*Leutnant*), Signals.
35 Belt buckle of pattern worn by Army Lieutenants, Directors of Music (*Musikmeister*) and Administration Officials accorded equivalent rank.
36 Brocade dress belt, *Luftwaffe*; for ranks as (35).
37 Brocade dress belt with gold buckle, worn with parade uniform by Wehrmacht Generals and Administration Officials of equivalent rank.
38-39 Shoulder straps, General of Cavalry (*General der Kavallerie*). The gold-yellow *Waffenfarbe* is irregular: officially, Army Generals wore bright-red *Waffenfarbe* and Luftwaffe Generals silver. But "rank has its privileges"—a Cavalry Colonel promoted to General might elect to keep his arm's *Waffenfarbe*.
40 Gilt aiguillette (*Achselband*), worn around right shoulder and across right breast (buttoning beneath right shoulder strap and on second and third tunic button) on parade and full dress uniform by Army Generals and Field Marshals and Administration Officials of equivalent rank.

Insignia of the Armed Forces of the German Federal Republic (Bundeswehr). *Left to right; top to bottom: formation badges of Ordnance Depot* (Materialamt des Heeres), *1st Airborne Div, Military District Command IV* (Wehrbereichskommando IV); *Luftwaffe Specialist's insignia; Officer's belt buckle, Army and Luftwaffe; Pilot's breast badge, all services, with gold oakleaves signifying more than 2,000 hours' flying time; Luftwaffe breast insignia; Army identification disc; Name tag for fatigue dress; Cuff title (obsolete), Army NCO School; Cuff title, Luftwaffe, now replaced by breast insignia shown above—a similar cuff title, with a grey cloth foundation, is still worn by the personnel of the Army Flying Corps* (Heeresflieger).

1 Staff Car Pennant, 1st Canadian Army, World War II. This cosmopolitan unit, with Czech and Belgian armoured and Dutch infantry brigades, distinguished itself at Falaise in August 1944.
2 Armband, worn on the left arm, by members of Military Police units of the US Army logistical organisation U.S. TASCOM, formed in France in April 1953—hence the *fleurs-de-lis* insignia.
3 Military Police armband, British Army, World War II. This armband was worn at a time when the unit's title was the Corps of Military Police; in 1946, the often-reviled "Redcaps" (see *Inset*) became the Corps of Royal Military Police.
4 Armband of an Imperial Japanese Army unit, World War II.
5 This armband of the Nazi Party's youth organisation dates from 1924, when it was a section of the *Sturmabteilungen* (SA, "Brownshirts"). After 1933, all German youth organisations were swallowed up by what became, in 1935, the *Hitler Jugend* (HJ, "Hitler Youth"), membership of which soon became compulsory for all children over the age of ten.
6 The tricolour armband of the French Forces of the Interior (FFI), the military branch of the French Resistance of World War II, trained in remote parts of France to assist the Allied invasion of Europe in 1944.
7 German armband of World War II, with official authorisation stamp to the left, worn by workers pressed into service for winter work (snow clearance and recovery after Allied air raids) in the Hanover district of north Germany.
8 Armband of the *Deutscher Volkssturm,* the last-ditch "Home Guard" organisation formed in Germany in 1944; see also (9) on *pages 166-167.*
9 Armband worn by the unsung heroines of the British "Home Front" in World War II—the Women's Voluntary Service (WVS). Formed in 1938 and with some 5,250,000 members by 1945, the WVS's various activities included helping with the evacuation of children, re-housing air-raid victims, canteens and National Savings schemes.
10 Utilitarian armband to be worn with an overall by such people as canteen workers of the British Women's Royal Naval Service (WRNS, "Wrens"), World War II.
11 This cap tally (the band worn around the hat) of the Women's Royal Naval Service (WRNS) dates from the corps' foundation in World War I.
12 Armband worn by a long-serving member of the British Women's Land Army (WLA), founded in June 1939 to free men from agricultural work for the fighting forces and with a strength of more than 100,000 by 1945 (see also 20 and 21).
13 Printed on felt, this badge of the Royal Canadian Engineers dates from the World War II period. It was possibly a mess decoration.
14 Locally-made armband of a Sussex branch of the British Local Defence Volunteers (LDV), fore-runners of the Home Guard,

May-August 1940.

15 Car Pennant of a Nazi Party member, Gemany, 1933-45; membership was estimated at around 6,500,000 in 1943.

16 Even the armband worn by German Red Cross workers in World War II bore the Nazi swastika. The design of the eagle varied from region to region; the one shown is probably a Croatian district.

17 The ubiquitous Nazi eagle; here in the form approved for wear on sports' singlets.

Other Rank's Service Cap, Corps of Royal Military Police; with the cover that gave the British Army's policemen the nickname "Red-caps". This example bears a post-1952, Queen Elizabeth II pattern, anodised badge.

18-19 Armband and (19) Car Pennant, the latter for a high-ranking officer, of the military branch of the *Sturmabteilungen* (SA), the brown-shirted "Storm Troopers" formed in the 1920s as bodyguard for the Nazi Party's leaders.

20-21 Armband of an early pattern and (21) red armband awarded after four years' service, for the Women's Land Army, see (12).

22 Armband issued in World War I to volunteers for the British Army to whom, because of shortages caused by a massive influx of recruits, the issue of uniforms and equipment was delayed.

23 Armband of a Nazi Party member (see also 15). Membership was not compulsory, but all those in official positions – down to the lowest level – were required to join.

Firearms

Terry Gander

Above: *US Marines assault a Japanese-held position on Okinawa, May 1945; the Marine in the foreground holds a ·30 calibre M1 carbine, while the Marine immediately to his left takes aim with a ·30 calibre Garand M1 rifle.*

Left: *Pakistan's North-West Frontier has long been a centre for the manufacture of reproduction firearms. Here, a craftsman admires a completed Lee-Enfield rifle.*

Right: *King's African Rifles, Kenya, 1950s: the FN rifle (left) was combat-tested in the Mau-Mau emergency; the man on the right has a Sterling L2A3 sub-machine gun.*

Far right: *A US soldier cleans a Smith & Wesson ·38in revolver; Pacific, 1944. Many firearms' enthusiasts have begun collections with an arm of this type.*

Firearms are a collector's dream, for they convey all the things that the collector wants. They have an attraction that is often coupled with an intrinsic value that will not normally fall or even decline, since they are produced in finite numbers when compared with other militaria. They are produced in an array of models, marks and sub-marks that allow the collector to look continually for that variation or addition that makes collecting the worthwhile occupation it is.

GUNS AND THE LAW

Having said this, the question arises as to why so few people collect firearms. The immediate answer is that in most countries today owning a firearm is very difficult. To have a firearm without official sanction is against the law, but to be within the law is sometimes made very difficult. At every turn of gun collecting the individual comes into contact with the law and with the police forces who have the unenviable task of administering the complex laws affecting the ownership of firearms. Although the summary that follows is based on the position in the United Kingdom, the laws of most European nations, and of some states of the USA, are equally complex and severe.

The law of the United Kingdom ensures that anyone infringing the Firearms Acts is severely punished, and for this we have to thank the increasingly violent times that we live in. Firearms are now commonly used in crimes against persons and property, as well as in acts of terrorism. Collectors thus find themselves restricted in their pastime, and the novice who wishes to start a modest collecion must note several points before he even thinks of purchasing a firearm of any kind. First, check the provisions of the various Firearms Acts currently in force. Ignorance of the law is no excuse for non-compliance, so be sure of what the law is and try to find out exactly how it is administered in your particular area. The latter point is important, for the wording of some of the Firearms Acts is rather woolly and it may be interpreted in ways that suit local Chief Constables rather than as they were drafted. In some parts of the United Kingdom, police forces are understanding and most helpful to collectors; in others they seem to be uncooperative to a degree. So do not expect the police to provide you with all the necessary details: not only are they always busy with graver matters, but in some areas they are under instructions from above actively to discourage the private ownership of firearms—espcially by collectors, who they seem to regard as people who are making their lives difficult for no real reason.

A few "don'ts": don't attempt to

obtain a firearms licence for an automatic weapon (ie, one that will fire more than one shot for a single pull of the trigger), for the ownership of *any* automatic weapon is prohibited, and that is that. Don't try to obtain a firearms certificate if you have any kind of criminal record: in 99 cases out of 100 it will not be granted. Don't obtain a firearm *before* applying for permission and doing the paperwork, or you will find yourself in serious trouble and so will the individual from whom you obtained the weapon.

With the obvious "don'ts" (there are others) out of the way, the time has come to see exactly what the would-be collector *can* do to own a firearm. In very simple terms, there are three ways to own a firearm of the rifle type: one may purchase a deactivated weapon; a smooth-bored example of a rifled weapon; or an unadulterated rifle.

WHAT KIND OF WEAPON?

The first option, the deactivated weapon, is attractive for those who simply want something to hang on a wall, or who wish to have a weapon to handle only. The main point about a deactivated weapon is that as it cannot be made to fire no firearms certificate is required and neither is any form of police permission. But a rifle that

cannot fire is just a piece of metal, not a firearm, so the collector will normally avoid such an expensive alternative. The word expensive is used advisedly: any deactivated firearm has to be deactivated from a complete weapon, and that weapon has to be paid for at full market price before the cost of the deactivation. In some cases deactivation can add up to 25 per cent to the market price. It is simple to purchase deactivated weapons of many kinds, but the degree of deactivation will vary from dealer to dealer. For some, deactivation means welding bolts closed, cutting out visible chunks of the barrel or some other mutilation. To some people, this degree of damage will be quite acceptable, especially if the weapon is to be merely hung upon a wall, but to the collector such damage is unacceptable to a high degree.

The next option is the smooth-bored rifle: a rifled weapon converted into a shotgun by boring out the rifling to a standard shot-gun calibre, usually, but not always, ·410. To the purist such a conversion is desecration, but it does mean that a collector can have an example of a particular weapon without too much trouble: shotguns require a shotgun certificate, but the current state of the law is that although a shotgun certificate has to be obtained to purchase a shotgun, any person can obtain a certificate unless the police can determine a reason why they should not have one. The law states that a shotgun is any weapon with a smooth-bored barrel not less than 24in (609·6mm) long, and rifles with smooth-bored barrels fall into that category. But the correct application forms must be made out truthfully and in full, and although the law does not at present make any demands as to security of the shotgun once purchased, the individual would be well advised to treat any weapon as though it were an item demanding the highest storage security.

While the adoption of the shotgun alternative may be a way for some

collectors to obtain a rifle, they may not like having to go through the expensive process of having the barrel bored out and then subjected to the legally-required proofing process at additional cost. And, of course, a smooth-bored barrel removes the weapon from its "authentic" state. The true collector will always want a weapon in as near its original state as possible —but if this is to be accomplished the collector will have to apply for a full firearms certificate, and this is where problems arise.

As the law stands, the would-be collector must convince the legal authorities (ie, the police) that he should be granted a firearms certificate for collecting purposes. Some police forces will need a lot of persuasion: the onus is on the applicant to convince the police that he is a genuine collector—and if the police refuse to grant a certificate there is little the would-be collector can do to reverse the decision. The form involved in applying for a firearms certificate asks for what purpose the firearm is to be used: typical categories include target shooting, pest control, sporting purposes, collecting and any other purposes. The main point here is that if a *collector's licence* is requested, and if it is granted, no ammunition-holding permission will be granted to go with the firearm—and the ownership of even a single round of live ammunition without a police permit is an offence against the Firearms Acts. So if you want to own a collector's firearm that can be *fired* legally, this will have to be stated on application. It is very unlikely that a certificate will be issued for than one rifle at a time, and the quantity of ammunition involved will be strictly limited. A collector's certificate may allow the holding of more than one firearm, but no ammunition.

SECURE STORAGE

Once the certificate has been granted, the problems do not cease. The police will insist on the weapon(s) being kept in a very secure place and may even specify the conditions of storage. Even collector's firearms usually have to be kept in a locked metal cabinet, chained or bolted to a fixed structure such as a chimney breast; such added conditions as keeping rifle bolts or other important parts in a separate locked cabinet may be imposed. The Police may call to check the certificate conditions and security at any time.

These extreme conditions upon ownership can be explained as being in the best public interest, but to many would-be collectors they are somewhat daunting. But there are ways in which the authorities can be persuaded that the would-be collector is a genuine gun enthusiast, a responsible person who wants to own firearms for his enjoyment only. One is to join a rifle or pistol club and take an active part in its activities; this is in itself a worthwhile activity and one which automatically gives access to firearms of all types. The club will not only provide contact with like-minded persons who will impart advice and guidance, it will also instil the disciplines that any form of firearms' ownership or use requires. Membership and regular attendance will often help in persuading the authorities that your enthusiasm is genuine, especially when hand guns are involved. The legal outlines given above relate as much to hand guns as to rifles, although a quirk of the law may allow the ownership of more than one pistol with the corresponding ammunition for certain types of sport and target shooting.

Another factor that might assist any application for a certificate is proof that the applicant is engaged in some form of research into various aspects of firearms' development or technicalities. What form this might take is up to the individual: it is a suggestion only. A further suggestion, that might allow some of the more esoteric modern rifles to gain acceptance on a certificate, is to take up the booming sport of "practical shooting"; firing at combat-type targets on what are often virtually combat ranges. This sport requires strict fire control and personal disciplines, with a fair level of skill, and is not for the dabbler. There is a "practical pistol" equivalent, involving the use of modern combat hand guns.

Having summarised the legal aspects of collecting firearms, the next important point is the type of weapon to purchase. This depends on the individual, and if the genuine would-be collector has done his research properly he will know what he wants. Proper preparation for ownership will require considerable research, involving the purchase of the relevant literature, visits to museums and collections, association with already-established collectors, and more. Firearms' collecting has the happy attribute that it need not be carried on by an individual in isolation: firearms' collectors are on the whole a gregarious bunch who like nothing better than to meet to swap information and guidance and generally talk firearms. There are collectors' clubs and associations to suit virtually any individual's field, and the association may be by meetings, postal exchanges or newsletters. By the time the collector has passed through this introductory process, he should have a clear idea of exactly what aspect of the hobby he is going to undertake, before purchase.

PURCHASING A FIREARM

At the purchase stage the would-be collector has another advantage not always shared by collectors of other militaria. That is, that in nearly every case the dealers involved are firearms' enthusiasts to a high degree. Their experience, knowledge and advice will usually be freely imparted. It is generally best to deal with trade members in the early stages of collecting, no matter how tempting some offers from private sources may appear to be: dealers have their reputations to maintain and will not want an unhappy purchaser on their hands. The golden rule, at all stages of collecting, is never to consider the acquisition of any firearm that is not "on a ticket", ie, one that is outside the legal system of ownership. Only trouble, possibly leading to the withdrawal of the firearms certificate and the confiscation of the collection, at least, will be the result of irregularities.

A man of the Italian contingent in the China Relief Expedition of 1900, sent to raise the siege of the Pekin legations by the "Boxer" insurgents, is armed with a 6·5mm Mannlicher-Carcano M.1891 rifle.

When purchasing a firearm there are many points to watch for. Since the market is finite, the price of an item alone will often give a fair indication of its collectable value—altough anyone who collects such less popular items as Webley service pistols or the readily-available Lee-Enfield rifles can still pick up real rarities for very reasonable sums. This is another of the delights of firearms' collecting: one collector's rubbish is another collector's gem, and a tyro collector who carefully chooses his field can have all the enjoyment of the pastime at moderate cost.

Whatever the price, there are obvious pitfalls to avoid. Always look for a specimen in as good a condition as can be found. This usually means one in which the internal and external metal is clean and free from pitting or old rust and wear marks. In some of the more elderly firearms these may be unavoidable, and in very rare pieces they will have to be accepted. As a general rule, the arm will bear markings of some kind, usually the place of manufacture and a trade or maker's name. Sometimes, especially on German firearms of World War II, these will be in code form and some reading and research will be required to understand them. There will usually be at least one serial number on any weapon, and on some of the finer models this serial number will be on every sub-assembly or part, so a quick check to ensure the firearm is still complete with its original parts is worthwhile. Some of the more involved collectors pay a great deal of attention to these serial numbers, for they alone may indicate a degree of rarity—some serial number sequences may denote a sub-contract with a small production run, for example—and in this way collection-enhancing specimens may be obtained by the well-read researcher from less well-informed vendors. Some collectors specialise in low serial numbers or end of production runs only—but their numbers are few and they tend to have a great deal to spend.

BEWARE OF COPIES

Whatever the weapon, its markings should be clear, legible and correct. The latter point is particularly important if Spanish or other copies are involved, for throughout the history of firearms' production, design plagiarism and the cheaply-made copy have been present. Spain has not monopolised the low-cost copy field; nearly every arms-producing country has had its opportunists at one time or another. Some copies are correct down to the serial number and product name and manufacturer, so the buyer must always be cautious. Beware of what purport to be established designs which show too many markings or excessive wear; these are just two danger signals to watch for. Once again, careful research, the advice of experienced collectors, and purchase from established firearms' dealers will cut the risks. This said it should be added that some collectors delight in the collection of fakes and copies!

An introduction of this kind can only serve to provide a very broad indication of the richness of firearms' collecting—but if one point can be established, it is that firearms' collecting is an absorbing and worthwhile occupation, but one in which the law imposes strict requirements of ownership and conduct. The ownership and use of firearms must involve strict self-discipline on the part of the individual. If this self-discipline is lacking, the ownership of firearms can have no good results. The collector who lacks it will not persevere with the necessary reading and research, nor will he carry out the security precautions that ownership entails; to mention just two serious failings. Such undisciplined conduct will soon become apparent to others in the collecting field, arousing suspicions that will rob the individual concerned of one of the most rewarding aspects of firearms' collecting: the social gathering together of like-minded people who enjoy nothing better than to meet and "talk guns".

Ammunition

Mike Priest

Ammunition may be collected in two states: live, with all its components as manufactured; inert, with the propellent powder removed and the primer fired or render inert.

To collect live rounds, a Part I Firearms Certificate must be obtained from the local police force (in the UK; similarly strict laws for the control of live ammunition are in force in most European countries and in some states in the USA). As with firearms, the police will require adequate reassurance and documentation before a certificate is granted. If granted, it will allow a specified number of rounds of a specified calibre to be held. While this will allow the collector to acquire variations on a single type, such as ·303 calibre, he may find it rather frustrating to be allowed to hold *x* number of rounds of a particular calibre and then to find a further variation which makes the number equal *x-plus-one!*

It is easier, cheaper and safer to collect inert ammunition. For this, no certificate or licence is needed, and a worthwhile collection can be started, with a good cross-section of 20 or 30 cartridges in basic calibres, for no more than the price of a meal in an average restaurant. One may choose to assemble a collection of rounds of a particular calibre; rounds for a single type of arm; rounds of any one country of origin (which may not be the same as the country of manufacture); or some other thematic arrangement.

A round of ammunition is made up of four basic components: the head, or projectile; the case; the propellant; the primer. (In inert rounds, as noted, the propellant will have been removed and the primer either fired or made inactive.) Heads are found in a variety of shapes and materials: as solid lead, or with a metal outer case, or jacket, of copper or nickel; round-nosed, pointed, hollow-pointed. Cases are most commonly of brass, but may be steel, chrome-plated, aluminium or plastic and aluminium (the metal forming the base); their shapes vary according to the chamber and bore of the weapon for which they were designed, including parallel cases, necked-down, tapered, and rimmed and rimless.

The various types of propellant include black powder, cordite and nitrocellulose; black powder is normally found as granules, cordite in strips or flakes, and nitro as cord, single-perforated or multi-perforated. Primers, the small explosive devices that fire the propellant, also come in several types and sizes, including the Boxer, Berdan and electrically-fired primers.

At this point, a warning must be given. *Do not attempt to tamper with live ammunition.* Do not remove the head; do not interfere with the primer; do not experiment with the propellant. The reasons are obvious!

HEADSTAMPS AND COLOURS

The summary above refers generally to small arms ammunition, but the principles of the fixed round extend to cannon and artillery rounds also. Some collectors favour large-calibre shells, since a good display can be made with a small number. Most often, you will find a large shell case without its fuse, or head—or the head without the case—but some specialist dealers handle large-calibre dummy rounds.

Once you have established a collection of basic calibres, you may increase its interest by a study of headstamps. Found on the base of the cartridge, these are like fingerprints: from them, you can learn the manufacturer, type and date of the round, and it is possible to assemble a fascinating collection of rounds of a single type, such as the ubiquitous ·303, all with different headstamps. Just one example of a headstamp may be given here (many more are listed on *pages 156-157*): "RG ⊕ 80". Here, "RG" shows that the manufacturer is the British company Radway Green; " ⊕ " denotes Nato ball type; "80" shows the year of manufacture as 1980.

Various colours will also be found on rounds of ammunition. The colour codes used vary from country to country, although there has lately been a move towards some standardisation. On some rounds, the primers are coloured to denote the type of bullet in the case: in Britain, for example, purple for ball ammunition, red for tracer (and in the latter case, the tips of the bullets are also coloured red). Colour codes may change with period as well as country: for example, in World War II, US military ·30 calibre tracer was marked in both red and white, but after the War orange was the colour most commonly used.

Finally, going beyond the rounds themselves, it may be noted that ancillary items like ammunition stripper clips, links, magazines and pouches (see *pages 158-159, 160-161*) make a good collection in their own right. Again, there is a wide variety of different types to collect. Most modern weapons accept more than one size or type of magazine; pouches vary in design from country to country; stripper clips come in varying sizes and links in different materials.

Above: *Boer General Christiaan de Wet and his staff carry their Mauser ammunition in bandoliers; South Africa, c1900.*

Left: *Corregidor, 1945; a US paratrooper feeds a distintegrating-link belt of ·30 calibre ammunition to a light machine gun.*

Service Revolvers of World War I

1 Gasser 11mm Montenegrin revolver. These massive arms were made in Austria, where the last model appeared in the 1890s, and in Belgium before World War I, mainly for the Balkan market; hence the name. There were many different models and barrel lengths, but all were huge and heavy. These revolvers are not rare: condition is an important factor when considering a purchase.

2 Smith & Wesson ·44 Russian Model revolver, produced 1870-78 for the Russian Army and then commercially as the No 3 New Model. There are many variants of this classic military arm, which is eagerly sought by collectors. Because of their often hard service, examples in good condition are very rare and expensive.

3 Smith & Wesson ·32 Safety revolver, originally made as a civilian pocket pistol in 1888 and produced in three versions, all with grip safety and shrouded hammer. Many found their way into service use and the type makes a good "study piece" on which to base a collection. Be wary of cheap specimens: the type was widely copied in Spain and Belgium.

4 Chamelot-Delvigne 10·35mm M1872 revolver, made largely for the Italian Army—hence the odd calibre, and in many ways typical of the enormous Belgian production of the late 19th century. Good examples are rare, but poor specimens are cheap: look for one with clear markings.

5 Nagant 7·62mm Model 1895 "gas-seal" revolver, made in large numbers for the Russian Army but not often seen in the West. It is unusual for the "gas-seal" action of its forward-cammed cylinder and for its odd ammunition, with the bullet inside the cartridge case. It is found in two forms: double-action, for officers; single-action, for ORs. Well worth seeking out, although condition may be poor.

(Top) Colt ·45 New Service; in US service from 1907. (Above centre) Webley & Scott ·455 Mark VI; ·22 LR version, with a ·22 barrel held within the ·455 barrel by a muzzle-screw, for training. (Below centre) Webley & Scott ·455 Mark VI; standard in the British Army, 1915-36. (Bottom) Swiss Ordnance 7·5mm M1882; Swiss service from 1882.

6 Pistole Revolveur Modele 1892. Often called the "Lebel", this 8mm revolver was made by Spanish and Belgian firms as well as French State manufacturers, and there are many variations in detail. It is not rare—many remained in service use in World War II—but most examples are ex-souvenirs and in a poor state. Examples made in France are most valued: beware of some Spanish specimens.

7 Colt ·38 New Army revolver, a fine arm that should be in any good collection. It differs from the usual Colt designs in having a cylinder that swings out to the left. Produced from 1908, it is not uncommon, but it is likely to be expensive when an example is found in good condition.

8 Webley ·455 Mark V; Webley ·455 service revolvers were produced in a long series of models and sub-derivatives, and would make a fascinating collection alone. Most could be easily obtained at fairly low cost, but some, like the Mark V of 1913, shown here, are rare. Buy carefully: avoid badly-marked specimens and those in poor condition. Like all Webleys, this was built to last.

9 Rast and Gasser 8mm Austrian Service revolver. Austrian arms of this type were produced from *c*1880 onward for purchase by Austro-Hungarian Army officers. The one shown is the Model 1898 in 8mm calibre, which was adopted by the Italian Army and is fairly common. The earlier, larger-calibre models—including one in 11·3mm—are preferred, but are less common: they were often issued for naval service and suffered rough usage, and well-finished, more ornate specimens, some of which are found in 9mm, fetch high prices in good condition.

10 Trocaola ·45 Service revolver. Spanish revolvers vary in quality from superb to dreadful: this example was good enough for British Army purchase in 1915, and specimens may still be found at reasonable prices. Beware of others: many are poor copies of more successful designs, in many calibres, and the collector should not try to fire them without very careful checking.

11 Colt New Service revolver; in production from 1897 until 1944 and thus still available in great numbers. Designed for service use, it was made in many calibres and barrel lengths—the example shown is in ·45 calibre with a 7·5in (190mm) barrel—and it is difficult to make a general statement regarding relative values. Condition will be the most important factor when buying—and look for US or British military markings.

12 Webley ·38 Mark III revolver. This is a rather unusual arm: a Webley & Scott Mark III in ·38 rather than the usual ·455. But such oddities abound in Webley production and this "funny"—a police rather than a service arm although it dates from the 1890s and many were subsequently pressed into military service—is highly collectable.

Service Revolvers and Pistols of World Wars I and II

1 Mauser 9mm Model 1912 self-loading pistol (SLP, "automatic"). This series originated with the C/96 and many types followed, notably the C/98; many copies were made in the Far East. The Model 1912, or C/12, shown with its wooden holster that serves as a butt for long-range firing, was originally made in 7·63mm but, as the "9" incised on its butt signifies, was altered to 9mm Parabellum. Unconverted examples have greater value—but the "Bolo" and "Broomhandle" Mausers are so greatly prized by collectors that they are today pieces for the very rich collector only.

2 Colt ·45 M1911; a very large SLP, weighing some 39oz (1·1kg). The Colt M1911 and M1911A1, with many variations, are both still in military service and are fairly easy to obtain, as are spare parts, but they are expensive when found in good condition.

3 Webley ·455/·476 Mark I revolver; the first of the big man-stopping Webleys—rather clumsy and hard to handle, but solid and strong—produced, in 1880, for the British Army's use in colonial wars. An attractive collector's piece, and not difficult to find.

4 Fabrique National Browning Model 1900. This 7·65mm SLP was developed by its Belgian designers from a Browning patent and was ordered by the Belgian Army in 1900. Its major claim to fame is that it was the weapon used in the Sarajevo assassination in 1914. Many are still extant, albeit some are very worn, but beware of copies from the Far East: many were made there with even the Belgian markings copied.

(Top) Mauser 7·63mm C/96; first of a famous series—see (1) above—and one of the world's first successful SLPs. (Centre left) Astra 7·63mm Model 900; a copy, made by the Spanish firm of Unceta y Cia from the late 1920s, of the C/96, with an integral 20-round magazine (see loading clip above). (Centre right) Right side view of the famous "Luger" P '08, see (10) above. (Bottom) DWM 9mm Marine Model 04/08 one of the rarer "Luger" types, with two-position rear sight, used by the Imperial German Navy. Shown with a 1917 pack of 9mm Parabellum cartridges.

5 Beretta Model 1934/1935; these small SLPs, still in production, are easy to obtain, but the collector will desire an original military specimen, like the one here which is dated 1941. These wartime examples are found in 7·65mm, as here, or 9mm Short, and with either Italian or German markings on the receiver. They will fetch a fair price in good condition, but beware of well-worn specimens.

6 DWM Parabellum Long Pistol '08, the "Artillery" model of the famous Luger; shown with its stock/holster and 32-round "snail" magazine, the latter a collector's piece in its own right. Like the Mauser Model 1912 (1), this pistol is now outside the price range of most collectors, although examples in less than good condition are still obtainable.

7 Mauser 7·65mm Model HSc, first produced for German forces, particularly the Luftwaffe and Navy, in 1940 and still in production. An attractive SLP and not too expensive, but look for one dated 1940-1944—the 1945 specimens are of lower quality—and beware of post-war French examples. Examples in 9mm Short and ·22 calibre are scarce.

8 9mm Repetierpistole Model 1912, often called the "Steyr-Hahn". This Austrian pistol, widely used by the Austro-Hungarian Army in World War I, was made for use only with the 9mm Steyr cartridge, but some later examples were re-chambered for the 9mm Parabellum round—so care must be taken with ammunition if it is to be fired. A charger clip must be used to load, since the magazine cannot be removed.

9 Glisenti 9mm Model 1910; a small Italian pistol used in both World Wars and fairly common. One to be collected rather than fired: it uses a 9mm cartridge far less powerful than the usual 9mm Parabellum, so care must be taken with ammunition, and any degree of wear will make this Italian pistol unsafe to fire.

10 DWM 9mm P '08, universally known as the "Luger", and one of the most famous of all SLPs—so prices tend to be high. However, it is still found in large numbers: look for examples date-stamped up to 1917; Swiss examples; or specimens with markings other than German. Even P '08 accessories make a worthwhile collecting field.

11 Fabrique Nationale Browning Model 1910; yet another product of the profilic Belgian factory. The Model 1910 was first made in 7·65mm and later in 9mm Short also. Although well made it was not used by many armed forces—some were taken into German service—but saw widespread police use. Many Spanish and other copies have been made, so look for clear Belgian markings; the price should not be too high, but the collector will want an example that is still in the best possible condition.

Service Revolvers and Pistols of World War II

1 Russian State 7·62mm Tokarev TT-30; this SLP of 1930, made at the Tula Arsenal, is a rare item, and although many World-War-II specimens will be in poor condition the collector has little choice. There are several direct copies made elsewhere: look for the Soviet Cyrillic markings on the receiver.

2 Fabrique National 9mm Browning 1935, the "HP" or "Hi-Power"; still in production and in service with many nations, so examples are easy to find. Collectors favour the World-War-II Canadian production versions, made for China; these, which are rare, have Chinese markings on the receiver.

3 Bayonne 7·65mm MAB Modèle D; appearing in 1933 and still in production, this French SLP is also found in 9mm and ·22 calibres. Examples produced for German forces, 1940-44, are rather rare but eagerly sought after; these fire 9mm Short rather than 9mm Parabellum ammunition.

4 Japan State 8mm Type 94. First produced in 1934, this is a rare collector's item and desirable in any condition. Poorly-designed and often badly-made, it should not be fired unless checked by an expert beforehand.

5 Walther 9mm P-38; one of the most popular pistols with collectors and therefore expensive—although not hard to find, and still in modern production as the Walther P-1. This fine pistol, produced at many centres, is well-documented; so read some of the many works on the arm before making a selection. Note that war-time examples may be in poor condition, and examples dating from 1944-45 may be suspect as regards safety.

6 Beretta 9mm Model 1934; like all Beretta pistols, eagerly sought by both collectors and target-shooters, since it is beautifully-made and a joy to use. Made in several variants: those used by Italian forces before 1943—found in both 9mm Short and 7·65mm calibres—are generally regarded as the best. Good condition will increase the value considerably, since many wartime examples are now in poor condition.

7 Fabrique National 7·65mm Browning Model 1910; appeared in 1912 and still in production for commercial sale. It has been widely issued for service use and a version for the 9mm Short cartridge was made. Not rare: a fine pistol and a type on which to base a collection.

8 Smith & Wesson ·38/200 British Service revolver. Some 1,125,000 of these arms were made for the British and Commonwealth armies, largely in 1940-45, and it is likely to be the first weapon that many collectors will acquire. Its many minor variations make it an object of study in its own right. It is sturdy and handles well: a good starting weapon for shooters.

9 Fegyvergyar 7·65mm P-37; a somewhat simplified version of the Femaru-Fegyver Pisztoly 37M, produced originally for the Hungarian Army and then for German forces, mainly Luftwaffe,

in 1941-44. On of the most eagerly collected of all German service pistols—and now scarce, since most were used in Eastern Europe.

10 Webley & Scott ·32 Automatic; one of the many Webley SLPs produced in various calibres—a larger version of this type was made in ·455—before and during World War I. All are now rare: a collection of Webley SLPs would be interesting—but expensive.

11 Enfield ·38 Pistol, Revolver, No 2 Mark 1; this British revolver is far from rare—it is solid and built to last—and would be cheap enough to form the starting point of a collection. The Mark 1* version, without a hammer spur and intended for use in confined space, such as tank interiors, is now more frequently found.

12 Radom 9mm wz 35; originally made for the Polish Army and then, in 1939-45, for German forces as the 9mm P-35—the earlier examples being of superior quality and the later found in greater numbers. Examples with Waffen SS markings are rare.

13 7·65mm P-27 (t); another "German" service pistol, this one of Czechoslovakian origin as the Czech Army's 9mm vz 24. Made in large numbers, but Czech Army examples are now rare and some of the late German production is of less than good quality.

14 7·65mm Walther PPK; originally for Police use, sometimes in 9mm Short calibre. One of the finest small SLPs ever made, the PPK is still in production and can be obtained "mint"—but collectors pay high prices for World-War-II military specimens. Check markings carefully, for copies have been made in several countries, including France and Turkey. Note that the arm shown lacks the usual finger extension on the magazine base; there may be other variations in these pistols.

(Above left) Roth-Steyr 8mm Model 1907; first SLP adopted by a major Army, Austria-Hungary, 1907. (Below left) SIG 9mm Model P-210-2; in service with the Swiss Army, as SP 47/8, from 1948. (Above right) Walther 7·65mm Model PP; forerunner of the PPK, see(14) above. (Below right) Sauer 7·65mm Model 1913; recommended by German War Ministry in 1930s for the use of staff officers.

1 Rifle No 4 Mark 1, ·303; Britain. The British No 4 Mark 1 Lee-Enfield of World War II was the result of a series of trials that began with developed versions of the No 1 rifle of World War I. The No 4 had a heavier barrel and revised sights, and was designed from the outset to be made by modern methods of mass-production; the bulk of production was by Fazakerley, BSA, and the Long Branch arsenal in Canada (where the No 4 Mk 1* was produced using slightly revised methods). Others were made in the USA by Stevens-Savage. Like all Lee-Enfield rifles, the No 4 will be found well-provided with stamped-on markings to denote the place of manufacture. This first-class service rifle may be regarded as a collector's dream: there are enough minor changes and sub-variants to keep the most ardent gun buff happy and many collectable accessories. Sniper models, inert training models, cutaway models and drill models are not difficult to find. Some reading on the subject will be worthwhile for the collector before purchasing a No 4.

2 No 9 Mark 1 bayonet; used with the No 4 rifle, but produced after World War II and mainly issued to the Royal Navy.

3 Mosin-Nagant Carbine Model 1944, 7·62mm; Russia. The Red Army's Model 1944 was the last of a Mosin-Nagant series extending back to 1891, in which most models were long rifles with an overall length of around 48·5in (1232mm). The Model 1944, however, was a short carbine—40in (1016mm) overall—for use by gun-crews, truck-drivers and other second-line troops. All Mosin-Nagant rifles are uncommon in the West and are avidly snapped up when they appear on the market—and since the handy length of the Model 1944 kept it in Warsaw Pact service almost to the present day, the few examples found are often not in very good condition. Mosin-Nagant sniper outfits with telescopic sights are particularly rare, and very expensive.

Effective but cumbersome body armour and heavy stahlhelm of German Army snipers and others in exposed positions in the trenches during World War I.

4 7·92mm Kar 98k; Germany; the standard German service rifle of World War II. Although designated a carbine (Kar=Karabine), it is about 39·5in (1000mm) long overall and has a five-round magazine. Like its World-War-I equivalent, the 7·92mm Gew 98, the Kar 98k is of great interest to collectors and, since it was mass-produced in huge numbers, is relatively easy to obtain and not too expensive. As with all German military products of World War II, some knowledge of the code-stamps denoting the centre where any particular item was manufactured or assembled is necessary, while reading will disclose the many versions and variants that may be encountered. Some knowledge and care is needed to acquire a good example of the Kar 98k. Later in the war production standards fell and some niceties were omitted. Look for an example on which the bolt is easy to work and locks well; where the markings are clear; and where the wooden furniture has no damage—a particularly important point on later models, where wood of poor quality was often used. Make sure that such items as the cleaning rod (not fitted to the example shown here, but usually located under the muzzle) are present. Again, sniper rifles with telescopic sights have a much greater value.

5 Mannlicher-Carcano Carbine Model 1938, 6·5mm; one of the later models in a series dating from 1891. There was a great range of Mannlicher-Carcano arms, including some that differed only in such details as sling swivels or butt shapes to suit some branch of the Italian services. The Model 1938 shown is in 6·5mm but some can be found in 7·35mm—planned as the new Italian service calibre before 1941—and some, captured examples bored out by the Germans, in 7·92mm. Thus there is a bewildering mess of calibres, lengths and models, which means—although they are fairly easy to acquire and not very expensive—that many gun dealers prefer not to handle them. Some late production examples are very poorly made.

6 Springfield Model 1903, ·300; USA. Taking its popular name from the plant at which it was originally made, the Springfield is one of the most prized service rifles of World Wars I and II with collectors and shooters alike. The M1903 shown here was the earliest version; the M1903A3 was simpler all round; the M1903A4 was a sniper rifle with a telescope and no "iron" sights, and is considerably rarer than the other two. This version apart, the Springfield remained in service until 1945 and so many were made that they can still be obtained at fairly reasonable prices.

7 Although the bayonet shown here was sometimes used with the M1903 when it was first issued for service, the two prominent rivets on the grip show that it is the bayonet issued for the earlier Krag Jörgensen Model 1896 rifle.

Service Rifles of World Wars I and II and After

1 Garand M1 rifle, ·30; USA; the standard US service rifle of World War II and thus as desirable to the collector as the British No 4 and the German Kar 98k (see *pages 150-151*). The Garand M1 was originally calibered for the American 0·30 rifle cartridge, but some were re-chambered for the very similar NATO 7·62mm—which will cause problems if used with older rifles. The Garand is a complex weapon that requires some care: the powerful bolt action can damage unwary fingers. Accessories abound, ranging from charger clips (see *pages 160-161*) and slings to complete sniper outfits. There are many Garands on the market, in conditions varying from near-mint to battered, but the most eagerly sought (and costly) will be presentation arms like the one shown here: the plate indicates that it was presented to the British Army's School of Infantry at Warminster, Wiltshire (where this weapon was photographed), by the US Army's School of Infantry at Fort Benning, Ga., in 1952. Such unique presentation weapons must not be confused with the various commemorative weapons produced by some manufacturers, which have engraving or plates marking some special occasion. Although these too are very desirable collectors' pieces, they are not in the same class as presentation weapons like the rifle shown here.

2 Bayonet-knife M5A1, issued for use with the Garand M1. The M5A1 is one of a series worth collecting in its own right, with many variations.

3 Ruger Mini-14/20GB rifle, 5·56mm; USA; a military version of a rifle first produced, and still in production, for a commercial and police market. A very well made rifle—sometimes finished in stainless steel rather than "blued"—the Mini-14/20GB, fitted with a combined flash-hider and bayonet lug and with parts of the stock made of fibreglass rather than wood, is in service with police forces and with some of the smaller Third World armies. The non-military version is to be found in many gun clubs, where it is used for everything from target shooting to hunting.

4 SMLE No 1 Mark 111, ·303; the Short Magzine Lee-Enfield Mark III, or No 1 Mark 111, was the standard British and Commonwealth service rifle of World War I, and many remained in use until 1945, despite the introduction of the No 4 Mark 1 (see *pages 150-151*). Such is the wealth of variations and sub-marks that the SMLE is a study in itself, and many weapons buffs collect only Lee-Enfields. The No 1 was virtually handmade, in the UK, India and Australia: look out for Australian examples and those made at Ishapore, India; production in Australia continued until 1957. The main variant is the No 1

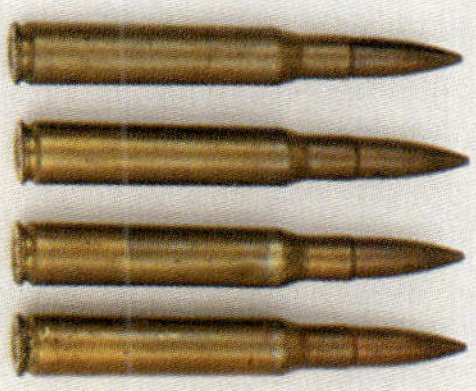

(Above left) 7·65mm P.Mod 37, Hungarian pistol produced for the Luftwaffe in 1941-42. (Below left) 7·65mm Ortgies, a German commercial design. (Above right) 9mm Browning 1935 "Hi-Power" made for the German Army, with backsight graduated to 550yds (500m). (Below right) 9mm Walther P-38 with silencer and modified sights, produced for Germany Army commando units, World War II.

Mark 111*, which has no cut-off plate below the bolt on the right-hand side, and there are ·22 versions, cutaways for training, drill versions, well-worn rifles adapted for firing grenades, and many more. There are still many No 1s on the market and they are not notably expensive, but some research and reading before buying may be worthwhile in both monetary terms and in terms of collectors' rarity.

5 A hinged trap in the butt plate of the No 1 rifle (and the later No 4) contained an oil bottle like the brass one seen here (later examples were in plastic) and pull-through; the latter for pulling "four-by-two" cleaning pads through the barrel.

6 Bayonet Pattern 1907; the type used with the No 1 rifle throughout World War I. This was one of a long series with the usual sub-variants. Very collectable and fairly easy to acquire.

7 Springfield M1903A4, ·30; USA. This is the special sniper version of the Springfield Model 1903 rifle (see *pages 150-151*); it has no conventional sights and relies on its Weaver telescopic sight alone. The general standard of these rifles was higher than that of ordinary versions and they were usually well cared for; however, although widely used in 1941-45, many were sold off for conversion to hunting rifles when sniper variants of the Garand M1 entered service and they are now comparatively rare. A complete M1903A4 outfit would today command a very high price.

1
2
3
4
5

1 Experimental L1A1 rifle, 7·62mm; Britain; a development model in the transition from the Belgian FN FAL assault rifle to the Enfield L1A1 service rifle. Such experimental models rarely come on the market; FALs and ex-Indian L1A1s—7·62mm arms firing single-shot only—are not particularly rare, although they are expensive.

2 Sauer Model 38H, 7·65mm; Germany. This self-loading pistol (SLP) was designed as a commercial pocket pistol, but most of the production—like the example shown—went to the German Army after 1939. Some of the late war models are of a lower standard than earlier production examples. The main attraction for the collector is that this is yet another German service pistol: examples with Wehrmacht acceptance markings will fetch considerable prices.

3 Armalite AR-15 rifle; 5·56mm USA. This was the commercial forerunner of the US Army's M16 and M16A1, and the receiver will be marked accordingly. It is in commercial production and may be purchased new, although it is very expensive. Further, the AR-15 has become associated with terrorist use, and its possession by private owners is viewed disapprovingly by the authorities. Automatic capability must be removed for civilian ownership, and even then a permit may not be granted. Since it fires the high-velocity 5·56mm round, special safety regulations may apply.

The German Army's first and current sub-machine guns: (above) Bergmann MP 18.1, 9mm, adopted in 1918; (below) the Israeli-designed UZI, 9mm, standard SMG of the modern Bundeswehr.

4 Star 9mm SLP; Spain. Unless he is a specialist in the field, Spanish pistols are a minefield for the collector: they vary from cheap copies to advanced and well-made models—and the same maker may turn out the best and worst of both categories. Most Star (Astra) designs are good and their production standards high: the example shown is in 9mm Parabellum. Bearing in mind the warning above, Spanish pistols are relatively cheap and may provide a good start to a collection.

5 Walther 7·65mm Model PP; Germany; also found in 9mm. A pistol popular with both collectors and shooters, still in production and relatively easy to obtain. Good examples will hold their value well, especially those produced for German forces before 1945 and bearing a service acceptance stamp. It has been made under licence in France, Turkey and elsewhere, and has been widely copied.

6 Russian State 9mm Makarov PM; little more than an enlarged copy of the Walther PP (5)—and thus a sound design. A few Soviet pistols come on the market from time to time, but their novelty value makes prices very high. Expense is increased by its ammunition: the Soviet 9mm Auto Pistol cartridge, a hybrid round rarely seen or used in the West.

7 AK-47 rifle, 7·62mm; the Soviet-designed AK-47 and AKM are unlikely ever to have scarcity value, for millions have been produced all over the world and are in service with both regular and irregular forces. Few come on to the collectors' market and they are very expensive, even when de-activated, as well as being subject to the same official restrictions as the Armalite (3). Research into origin and markings is a necessary preliminary to purchase: there are many variants and some countries of origin are rarer than others.

8 Remington ·45 M1911 SLP; a copy of the Colt M1911—see (2) on *pages 146-147*—and an example of how an ordinary pistol can become a collector's prize. In World War II, Remington made the Colt design under licence, but only about 13,000 were made and they have become rareties commanding high prices.

9 Developmental design for a bayonet for the L1A1 rifle (1); it was not accepted for service and is now very rare.

10 Colt ·32 Model 1903; not primarily a military pistol, although often carried as one, the Browning-designed Model 1903 is not particularly rare. The type has been widely copied: beware of "Spanish" examples.

11-11a Taurus ·357 Magnum revolver, Brazil; one for the pistol buff rather than the collector: it would be as well to defer collection of such pistols to a time when their market value has been better established. The special screwdriver-tool (11a) for the revolver is itself a rarity; most modern revolvers can be stripped without the use of tools.

Small Arms Ammunition of the World, 1900–1983

Countries are those in which the round was made; not necessarily country of origin. The headstamp of each round, in parenthesis, is included to give some idea of the complex fascination of this branch of collecting.

1 ·22 Short Dummy/Drill; UK (none).
2 ·22 LR Dummy/Drill; UK (none).
3 ·22 Short; USA; (Ⓡ).
4 ·22 LR; UK (E).
5 ·22 LR Tracer; UK (I.C.I.).
6 ·25 (6·35mm) Auto; USA (W-W 25 Auto).
7 ·32 Smith & Wesson Short Revolver; UK (none).
8 ·32 (7·65mm) Auto; Belgium (FN☆).
9 7·65mm Parabellum (7·65 Luger); Switzerland (T 9 T 65).
10 7·63mm (·30 Mauser); Italy (G.F.L. Mauser 7·63).
11 ·380 (9mm Short) Auto; USA (Western 380 Auto).
12 9mm Parabellum Drill; UK (none).
13 9mm Drill D Mk 2; UK (RG 72 9mm D2).
14 9mm Blank; Germany (Geco 9mm).
15 9mm Mk 2Z; UK (RG 74 9MM2Z⊕).
16 ·380 Ball Mk II Service Revolver; UK (R↑L 39 ·380 II).
17 ·38 Indoor Practice; USA (Speer ·38).
18 ·38 Special Shot Shell; Finland (Lapua ·38 SPL).
19 ·38 Special; Finland (Lapua ·38 SPL).
20 ·357 Magnum Shot Shell; USA (S & W ·357 MAG).
21 ·357 Magnum; USA (W-W Super ·357 Magnum).
22 ·44 Indoor Practice; USA (Speer ·44).
23 ·44 Smith & Wesson Special; USA (W-W 44 S & W SPL).
24 ·44 Remington Magnum Shot Shell; USA (W-W Super 44 Rem Mag).
25 ·44 Remington Magnum; USA (W-W Super 44 Rem Mag).
26 ·44-40 Winchester; UK (Kynoch 44.W).
27 ·45 ACP M1911 Ball; USA (R A 6 8).
28 ·45 Long Colt; USA (W-W 45 Colt).
29 ·45-70 Government; UK (Eley London).
30 ·455 Webley; UK (K42 V1Z).
31 ·22 Hornet; USA (W-W Super ·22 Hornet).
32 ·222 Remington; USA (R.P 222 Rem).
33 5·56mm (·223) Plastic Blank; Germany (DAG 5·56MM).
34 5·56mm (·223) Ball; USA (T W 7 2).
35 ·220 Swift; USA (Super-X 220 Swift).
36 ·22-250; Finland (Sako ·22-250).
37 ·243 Winchester; Sweden (Norma 243).
38 6·5 x 50mm Arisaka Type 38 Ball; Japan (none; plain).
39 6·5 x 54Rmm Mannlicher; Netherlands (10 25 25 A55).
40 ·240 Weatherby Magnum; USA (Weatherby ·240 Magnum).
41 ·264 Winchester Belted Magnum; USA (W-W Super 264 Win Mag).
42 ·255 Rook Rifle; UK (.Eley. ·255).
43 ·270 Winchester; USA (FC 270 Win).
44 7mm Rem Belted Magnum; Finland (.Sako. 7mm Rem. Mag.).
45 ·30 M1 Carbine; Belgium (F N☆).
46 ·30-30 Winchester; USA (F C 30 30-30 Win).
47 ·32 Winchester Special; USA (WRA.CO 32 W.S.).
48 ·30-06 Rifle Grenade Blank M3;

68 69 70 71 72 73 74 75 76 77 78 79 80 81 82 83

(Above left) Primary ejector-seat cartridge, UK; (below left) US 40mm Rifle Grenade for M79 Grenade Launcher; (centre) 1·5in (38mm) Anti-Riot Baton Round ("Rubber Bullet"); (right) 1·5in (38mm) Signal Cartridge.

84 85 86 87 88 89 90 91 92 93

USA (F.A. 41).
49 ·30-06 Blank M1909; USA (F.A. 34).
50 ·30-06 Dummy M1906; USA (F.A. 5 5).
51 ·30-06 Ball M2; USA (F.A. 32).
52 7·62mm Nato Blank L10A2; UK (RG 68 L10A2).
53 7·62mm Nato Drill L1A2; UK (RG 57 7·62 L1A2).
54 7·62mm Nato Drill; UK (RG 66 L2A2⊕).
55 7·62mm Nato Inspectors Round L3A1; UK (RG 56 7·62 L3A1).
56 7·62mm Proof Round; Australia (MF 76 7·62LAA1).
57 7·62mm Nato Ball; Norway (⊕14-RA-75).
58 ·303 Blank L Mk V; UK (R↑L 43 LⅤ).
59 ·303 Grenade Blank H Mk VIIZ (FN); Belgium (F N 52).
60 ·303 Drill D Mk V; UK (R↑L 15 IV).
61 ·303 Drill D Mk 10; UK (RG 55 D10).
62 ·303 Drill D Mk VIII; UK (none).
63 ·303 Inspectors U Mk V; UK (R↑L 1943 UⅤ).
64 ·303 Proof Q3; UK (RG 55 Q3).
65 ·303 Ball; Belgium (F N 50).
66 7·92 x 33mm Kurz Semi-Armour-Piercing; Germany (44 OXO St 3).
67 7·92 Mauser (Wood) Bulleted Blank; Germany (40 P25 VIIIg1 92).
68 7·92 Mauser (Plastic) Bulleted Blank; Spain (FNP 7·92-955).
69 7·92 Mauser Ball; Germany (38 P S* 60).
70 7·62 x 39mm Russian Short (CWB) case; USSR (539 73).
71 7·62 x 39mm Russian Short Steel case; USSR (3 70).
72 7·62 x 53Rmm Russian Mosin Nagant; Finland (Lapua 7·62 x 53R).
73 7·5 x 54mm French MAS M1929; France (I CN 34 PC).
74 7·5mm Schmidt Rubin; Switzerland (T 5 A 69).
75 7·65 x 54mm Turkish Mauser; Turkey (Arabic characters).
76 8 x 56mm Mannlicher-Schoenauer; USA (Western 8mm Mann-Schoen).
77 8mm Lebel (Wood) Bulleted Blank M1905-27; France (2 LM 39 LM).
78 8mm Lebel Model 1886D; France (4/AHT.D./10/V/S.R/).
79 8 x 60mm Magnum; Belgium (F N 8x60s).
80 ·300 Winchester Belted Magnum; USA (W-W Super 300 Win Mag).
81 ·300 Weatherby Belted Magnum; USA (Speer ·300 Weatherby Mag).
82 ·375 Belted Magnum; UK (Kynoch ·375 Magnum).
83 9·3 x 74Rmm Mauser; USA (9·3 x 74R Browning).
84 ·378 Weatherby Belted Magnum; USA (·378 Weatherby Magnum).
85 50 cal Spotter; USA (L C 77).
86 50 cal Browning Blank; UK (K 61.50).
87 50 cal Browning Dummy; USA (F A 4).
88 ·5 cal Browning M2 Ball; UK (K50.50).
89 ·55 Boys Anti-tank Practice Pl; UK (K 37 P.I.).
90 Primer "Tube" for 120mm Tank Round; UK (L1A4 RG 74 5N).
91 20mm Vulcan (M103A1) Steel Case; USA (none; plain).
92 20mm Vulcan Brass Case; USA (none; plain).
93 20mm Oerlikon Cannon Steel Case; USA (E.K. 1943 20MM Mk.2).

Ammunition, Clips and Belts, 1914-1983

1 Mark 2/2 bomb of the illumination type for the British Army's ML (Muzzle Loading) 2in Mortar, a weapon now being replaced by the 51mm Mortar. The bomb's filling comprises coloured illuminating stars and it is provided with a parachute for descent.
2 British-made blank round for use in training with the 40mm Bofors light anti-aircraft gun.
3 40mm Bofors gun round with fuse plug fitted in place of fuse.
4 Illuminating bomb L28A2 for the British ML 81mm L16 Mortar; like (1), this incorporates a parachute for descent. The 81mm L16 has replaced the 3in Mortar in British service and is under consideration for the US Army.
5 British-made Type 94 Energa practice rifle grenade fired with a grenade-launching attachment from the SMLE rifle.
6 7·65mm Turkish Mauser ammunition; the five rounds and stripper clip are of German make and date from World War I.
7 Ammunition on 10-round stripper-filler clip for US ·30 cal Carbine. The filler is inserted into the top of the magazine and the rounds are thumbed down; the stripper-filler is removed and the magazine inserted into the carbine.
8 ·223 cal (5·56mm) ammunition for the Belgian FN light machine gun system; on ammunition belt with metallic disintegrating links.
9 US Rifle Grenade Type M9A1; a grenade-launching attachment is used to fire this from the Garand M1 rifle, Springfield M1903 rifle and ·30 cal Carbine.
10 ·30 cal Tracer ammunition (note coloured tips) on 8-round clip for US Garand M1 rifle.
11 British-made shells for 30mm Aden cannon, an aircraft weapon. Note that five of the six rounds are in metallic links.
12 ·223 cal (5·56mm) ammunition for the US Army's M16 Armalite rifle, in 10-round stripper clip. Similar ammunition is now made and used in Britain.
13 French 8mm Lebel ammunition on three-round clip. Used until after World War I in the Carbine M1890, Mousquetoon M1892 and Rifles M1902, M1907 and M1907/15.
14 ·300in ammunition for Browning machine gun on a British-made 250-round fabric belt dating from the mid-1960s.
15 British-made ·303in ammunition on metallic disintegrating link machine gun belt. Belts and ammunition of this type were used for the wing-mounted Brownings of such World War II aircraft as the Supermarine Spitfire.
16 ·5 cal Browning M2 dummy ammunition; the four rounds, made with tin-coated steel cases and dated 1944, are shown on metal disintegrating links.
17 British 30mm AFV rounds for Rarden cannon (L21A1), in the 3-round loading clip used in the Fox armoured car. Two of the 3-round clips are inserted into the rear of the gun, the loader using the hand-grip at the base of the clip; the clip releases the rounds as soon as they are inserted.
18 British 7·62mm Nato blank rounds in disintegrating metallic belt for the

GPMG (general purpose machine gun). Note that they have been fired and the tips of the cases are blown open; they are normally tightly closed with bullet-shaped ends.

19 20-round loading strip for the Italian 7·92mm Breda Model 37 machine gun. It is interesting to note that after the rounds are fired they are replaced in the feed strip, rather than being thrown clear as in most automatic weapons.

20 7·62mm x 39 ammunition on 10-round stripper clip for the Soviet Simonov semi-automatic carbine (SKS).

21 ·303 Ball ammunition on 5-round stripper clip (or "charger"); the standard ammunition and clip used for the various British Short Magazine Lee Enfield (SMLE) rifles of World Wars I and II.

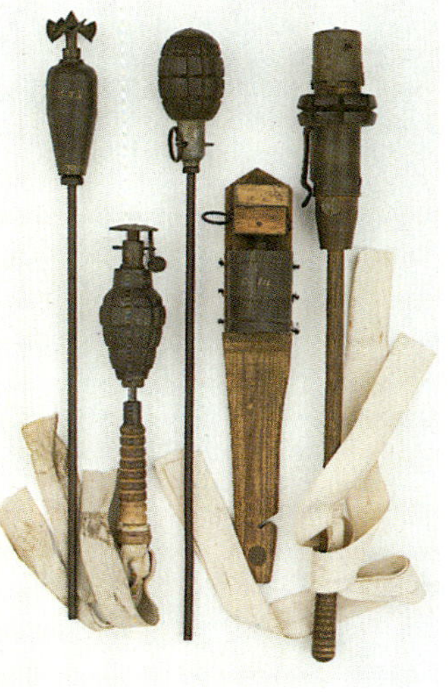

A line-up of British Army Rifle- and Hand-Grenades of World War I and later. From left to right: No 25 Rifle Grenade; No 19 Hand Grenade; Rifle Grenade variant of No 16 Hand Grenade; No 12 Hand Grenade (called, for reasons apparent in the photograph, the "Hairbrush Grenade"); Grenade, Hand, No 1 – the first official British hand grenade of World War I. Warning: always ensure that grenades, mortar bombs and similar specimens are inert; seek expert advice if necessary.

22 German machine gun belt of World War II with 7·92mm Wooden Bulleted Blank ammunition. This metal belt, formerly used with the MG42 and still in use by Germany with the MG3, is not of the individual disintegrating link type but comes only in 50-round lengths.

23 British-made ·5 cal Browning blanks on metal belt of disintegrating link type.

24 7·62mm Nato dummy rounds on 5-round stripper clip. These dummy rounds, British-made from cases dating from the late 1960s, consist simply of used cases with wooden spacer inserts and are used for drill and training.

25 British-made ·5 cal Browning ammunition in metal disintegrating link belt. Note the way in which the ammunition is ordered: this example shows four armour-piercing (AP) rounds, colour coded by the black tip, followed by one tracer (red tip); this arrangement would repeat over 100-round lengths all through the belt.

Ammunition, Magazines and Pouches, 1900-1983

1 32-round metal box magazine for US 9mm Parabellum Ingram M10 sub-machine gun.

2 20-round metal box magazine for US ·5 Spotting Rifle (Cal ·50 Spotting Gun M8C), the gas-operated SLR used as a ranging gun for the 106mm Recoilless Rifle M40. The ·5 round—note red and yellow identification code on nose—incorporates both tracer and incendiary elements, the latter emitting a puff of smoke on impact. Its trajectory is matched to that of the high-explosive anti-tank (HEAT) ammunition fired by the M40. The same rifle is used for ranging with the British Army's now-obsolescent Wombat 120mm Battalion Anti-Tank Gun, but with ammunition modified to match the trajectory of the larger-calibre British weapon.

3 34-round metal box magazine for 9mm x 19 Parabellum ammunition of British L2A3 sub-machine gun (Sterling).

4 10-round box magazine for British 9mm single-shot Police Carbine.

5 30-round metal box magazine for ·45 ACP ammunition of US M3 and M3A1 sub-machine guns ("Grease Gun"; so called because of its supposed resemblance to that piece of equipment).

6 20-round metal box magazine for 7·62mm Nato ammunition of British Self-Loading Rifle (SLR).

7 Ammunition loader for use with magazines of British Sten and Lanchester sub-machine guns (8). The loader is positioned atop the magazine and the ring-handle is pulled down to allow a cartridge to be inserted; the ring-handle is then raised to force the cartridge into the magazine. This allows magazines to be loaded to the fullest capacity.

8 32-round metal box magazine for 9mm Parabellum ammunition of British Sten and Lanchester sub-machine guns.

9 20-round metal box magazine for 7·62mm Nato ammunition of US M14/M14A1 rifle.

10 30-round transparent plastic magazine for 5·56mm x 45 ammunition of Austrian Steyr AUG (Army Universal Gun; carbine, assault rifle or sub-machine gun). A transparent magazine allows the user to check that he has loaded to full capacity and also on the number of rounds still unexpended when firing.

11 Pouch for US ·30 cal M1 Carbine magazines. The pouch, which holds two 15-round magazines, one in each pocket, is usually worn on the pistol belt but is sometimes attached to the butt of the carbine.

12 Pouch for 30-round magazines for US ·30 cal M2 Carbine. Worn on the pistol belt, this pouch will hold four curved ("banana") magazines.

13 15-round metal box magazine for US ·30 cal M1 Carbine; shown with 10-round stripper clip with integral magazine loader, for swift loading of 15- and 30-round magazines.

14 30-round metal magazine shown with the spring clip ("jungle clip") used for joining two magazines in "one up, one down" position, for swift accessibility of new magazine in action.

15 US Army issue ALICE magazine

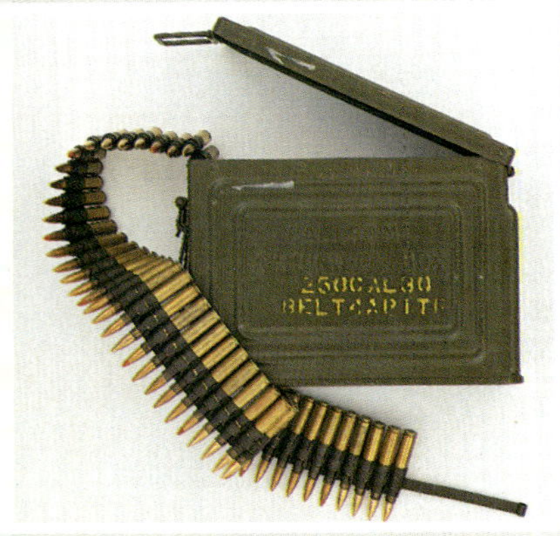

US M1 Ammunition Box for 250 rounds belted ·30 cal ammunition for Browning machine gun. The rounds are shown in a metallic link belt with a feeder tab.

pouch in nylon. The pouch holds three 30-round magazines for the ·223 cal M16 Armalite Rifle and two M61 ("Egg") high-explosive fragmentation grenades. Two of these pouches are worn on the combat belt.

16 8-round clip loader for ·30 cal M1 Garand Rifle, containing tracer ammunition (note orange tips). To the right are 8 rounds of ·30 cal Ball ammunition.

17 US Army ammunition belt with 10 pockets, holding a total of either 10 x 8-round clips for the ·30 cal M1 Garand Rifle or 20 x 5-round clips for the M1903 Springfield Rifle.

18 20-round magazine for US ·223 cal M16 Armalite Rifle, shown with 10-round stripper clip and magazine loader which, unlike the similar equipment shown at (13), are made in two separate pieces.

19 10-round metal box magazine for British Short Magazine Lee Enfield (SMLE) rifle.

20 30-round metal feed tray for French Hotchkiss Model 1909 light machine gun. The example shown is from the ·303in version used by the British—who often designated the weapon the Benét-Mercié Machine Rifle—in World War I; the French used it in 8mm Lebel. The tray or strip feed was preferred by the designers to the more common belt feed.

21 20-round metal box magazine for US ·30 cal Browning Automatic Rifle (BAR).

22 15-round box magazine for ·22LR Charter Arms AR7 (Survival Rifle); an 8-round capacity is more usual.

23 US-made 30-round extended metal box magazine for 9mm Browning 1935 (known as the "Hi-Power") self-loading pistol.

24 Magazine pouch for 2 x 7-round magazines for Colt ·45 M1911 self-loading pistol; see (27).

25 20-round stainless steel box magazine for 9mm Browning 1935 ("Hi-Power") SLP.

26 13-round stainless steel box magazine—the normal capacity, as opposed to the magazines of much greater capacity shown at (23) and (25)—for 9mm Browning 1935 ("Hi-Power") self-loading pistol.

27 7-round metal box magazine, shown with 7 rounds of ·45 ACP ammunition, for Colt ·45 SLP.

28 25-round extended metal box magazine for ·45 Auto ammunition for Colt ·45 self-loading pistol.

Printed Militaria

Terry Gander

Above: *This is how many personal documents began their journey to the present-day collector. German prisoners taken during the Normandy landings of June 1944 are searched by US soldiers. One offers his personal record book (see the* Wehrpass, *(left) and* Soldbücher *at (1) and (9),* pages 166-167*); many such documents were kept as souvenirs, rather than being turned over to the appropriate military authorities.*

Left: *German printed material of World War II: (left) Army manuals and a soldier's* Wehrpass *(Service Book); (central) Hermann Göring features in the* Berliner Illustrierte Zeitung, *1941; (right)* Signal, *the propaganda magazine published as much to influence the enemy as to boost German morale: note that one of the copies shown is an English-language edition.*

A modern nation at war generates an enormous amount of printed material, over and above the massive documentation produced by the armed forces in both war and peace. The militaria collector, whatever his specialisation, will almost certainly have the desire or the need to acquire some kinds of printed matter: training manuals and drill books for collectors of firearms and edged weapons; dress regulations for headgear and uniform specialists; rank lists and official gazettes for medals' researchers, and so on.

The general collector, as the newcomer is likely to be, has a wider choice. Wartime measures in any country involve the production on a grand scale of such ephemera as identity documents, ration books, special currency, information and propaganda material, recruiting leaflets, as well as newspapers, magazines and photographs. Leaving aside the collection of books (other than manuals) and military-based philately—both subjects that demand whole books in themselves—the field is one in which there is room both for the expert and the novice.

PRESERVATION

All the material illustrated on the following colour pages and discussed in this necessarily brief survey has one thing in common: it is printed on paper. For the collector, paper is thus both a source of fascination at its contents and worry as to its condition, for the collectables are only as permanent as the paper on which they are printed. The care and preservation of paper is an aspect that must be understood if the collecting of printed material is to be at all worthwhile.

Paper manufacture is often dependent upon imported materials, and for most nations in both World Wars this meant that the paper industry could not obtain such peacetime materials as wood-chippings and other timber-based products but relied on waste paper and other re-circulated waste, such as low-quality textiles. Paper produced from such materials is often of very poor quality and tends to crumble after an all-too-brief period. German paper products of World War II provide a good example of rapid "ageing": most of the surviving material appears to have been printed on fragile orange-coloured paper which is prone to crack and even fall apart when handled. Generally speaking, the Allied nations did not make such wide use of low-grade paper, but even so much Allied printed material is now also fragile.

Care and attention is demanded of the collector. Original sheets must be carefully stored in surroundings that will not affect them further. This need not mean air-conditioning: the average collector should simply store paper-based items away from extremes of heat, cold and moisture and out of the reach of insects. Moisture must be controlled: too much damp and paper will quickly deteriorate; but if the atmosphere is too dry it will crumble into dust. Complete lamination is not generally necessary unless the material is already breaking up. Smaller items may simply be kept in plastic wallets. For more bulky items, metal boxes are effective against paper-eating insects. Storage in cabinets or on shelves—at any rate, away from the floor where most paper-loving insects seem to live—is also effective; a regular routine of examination and dusting will then guard the collection against the more severe forms of insect depredation.

The Iron Cross, Second Class (1914-18), is by no means rare, but this example, awarded in 1918 to Gefreiter (Lance-Corporal) Karl Lausterer of the 10th Württemberg Infantry Regiment No 180, has been framed with its citation, bearing the official regimental stamp and signed by the regiment's commanding officer. A fine piece for the collector of printed material.

Even so, paper will inevitably deteriorate, and if its message is to be preserved the collector may eventually have no choice but to copy it. Researchers will generally wish to use a copy in any case; the collector will want the real thing—but a copy is better than nothing. Try to use a dry paper copying system, for most oxide or chemical based papers will fade with time. Alternatively, depending on the individual's resources, micro-filming or photography with a normal camera on a copy stand may be resorted to. The possessor of unique printed materials should always have them copied, whatever their condition, so that a record will remain in the event of any accident befalling the original documents.

MANUALS AND NEWSPRINT

The collection of military manuals and of wartime magazines and newspapers is now a very well-organised aspect of militaria collecting; most established collectors and dealers have a good idea of what is rare, what is common—and what it is worth. This need not discourage the novice—there is always room for a newcomer in any collecting activity—but it does mean that careful investigation is necessary before money is spent to begin a collection. As in any branch of collecting, time spent on research can prevent the novice from being landed with worthless or over-priced material.

Service manuals for weapons and some types of vehicles present one particular problem. Many service manuals for wartime vehicles are now so eagerly sought by "preservation buffs" for work on the restoration of vehicles that some commercial firms have found it worthwhile to produce facsimile editions. These are exactly the same as the originals in appearance and content, but the true collector will not want such copies. However, collectors are asked, and will pay, high prices for original manuals— particularly for weapons' manuals, which for many represent the nearest they will get to possessing the weapons themselves.

Service and training manuals are often most attractively produced, with coloured plates and well-annotated line drawings and photographs. The text will not only show how a piece of equipment worked or how a body of men was controlled: it will, if carefully read, provide an insight into the tactical and social attitudes prevalent at the time of issue. For example, manuals of World War I sometimes differ markedly from those of World War II, even when dealing with the same subjects: not only had tactics changed, but the general standard of education had improved. Further, manuals often contain hand-written notes or other inserts that convey the elusive sense of personal involvement that is so important to the true collector. And although many manuals are expensive, their incidence on the market cannot increase with the years and they should at least retain their value.

So far as newspapers and magazines are concerned, the most desirable items will obviously be those which cover the great events of a conflict. However, the fact that wartime newspapers still survive at all may be wondered at, since they were printed on low-quality paper and many were immediately disposed of in wartime waste-paper collection drives. Even rarer are copies of the special newspapers and news-sheets produced for troops in the field. Lucky finds may still be made, but the condition is likely to be poor. The same applies to propaganda leaflets, produced by the million during wartime, but now scarce and attracting fairly high prices. They are, however, most desirable collectables, since they were attractively designed to catch the eye.

EPHEMERA

Mention of relatively high prices need not deter the novice, for there are still items that can be found at little cost if the collector is prepared to take some time in searching. It is a worthwhile and absorbing activity to hunt down what are generally described as "ephemera". This category includes such items as civilian ration books and identity cards, soldiers' personal papers and letters, official photographs, occupation currencies, postcards, recruiting leaflets, Red Cross items, and a host of other things.

Ephemera may still be found in all manner of places, from house clearance sales to a few specialist dealers. Such items as postcards are already highly collectable: views of battlefields—such as the numerous post-1918 issues depicting the battlefields of France and Belgium—and special issues depicting important events are well worth looking for. Into the category of pictorial ephemera fall also collectables that the novice may be able to acquire from older members of his family: snap-shots taken by servicemen, unofficially, which often show aspects of military life neglected in the high-quality and widely-distributed official photographs.

Athough their material value may not be great, ephemera convey a sense of period, the atmosphere of time past, in a way that more valuable and concrete objects cannot. These humble collectables have a personal quality: they convey the important message that history, especially military history, involves people, not just machines and artifacts—a fact that the specialist collector may sometimes forget.

1 French Staff Manual on Heavy Artillery, World War I. Handbooks produced for staff officers are often summaries of the information that higher-ranking officers need to estimate the capabilities of the weapons under their command. A very similar edition was produced for use by students at the French Army's artillery school.

2 *The "Parabellum" Automatic Pistol:* an English-language sales handbook on the famous "Luger" self-loading pistol (see *pages 146-147*) produced by its makers, Deutsche Waffen- und Munitionsfabriken (DWM), Berlin. Such handbooks for prospective buyers are usually very well produced and illustrated and provide the weapons' buff with invaluable information; and some, like the example shown here, are considered good enough to be commercially reprinted today.

3 A modern Irish Army manual on the Carl Gustav anti-tank gun. As well as technical information on the weapon and its use, this manual provides details of drill and uniform that may be of great value to researchers in years to come: thus, today's manual becomes tomorrow's collectors' piece.

4 A Guide to German Ammunition, published by His Majesty's Stationery Office (HMSO) during World War II. HMSO often acted as a general publisher for the British armed forces in 1939-45, originating many handbooks containing general information as well as far more detailed works.

5 *TM 9-1900/TOA11-20 Ammunition General,* a combined US Army and US Air Force training and general manual relating to all types of ammunition and ordnance stores. A mine of information, such manuals are useful to researchers and modellers as well as to ammunition collectors.

6 Users' manual for the Oerlikon gunsight, issued by the US Navy. The 20mm Oerlikon predictor sight proved too complex for on-board use: few were issued, and the manuals are now all that survive of the type. Such handbooks on equipment that was never actually issued for general service use are rare and are well worth seeking out.

7 *Fusil-Ametrallador Vickers-Berthier,* a Portuguese-language sales handbook (see 2) produced by the firm of Vickers-Armstrong in 1929 for potential purchasers of the Vickers-Berthier light machine gun.

8 One number of a part-work on Italian artillery: a pre-1940 commercial publication sponsored by the Italian Army as a form of general artillery course for recruits and para-military personnel. The "students" were expected to purchase their own copies of the periodical; issues are not rare.

9 Although nothing more than a post-card sized piece of cardboard, this is a rare and unusual collectors' piece. It is a *Panzerschreck* target training card, issued to the German users of the 8·8cm RPzB 54 (a "bazooka" type anti-tank weapon) to show them the best points of aim on Allied AFVs of World War II.

10 Like (9), this is a "reminder" card

intended to be carried in the pocket or pack: a *Demolition Card*, a folded, laminated booklet, issued by the US Army. This example, *GTA 5-10-9*, was issued in 1969; many other dated examples may be found. Although relatively common, these are not easy to acquire because of the nature of the information they contain.

11 These two drill books of World War II are part of a long series issued by the German Luftwaffe for the use of "Flak" units using captured anti-aircraft guns after 1940-41. Now rare, these are extremely collectable because of the array of hand-stamps and other unit insignia their pages often display.

12 *Pistole 24(t)*, a German handbook of 1940 for a Czechoslovakian self-loading pistol. If you are unable to collect the weapons themselves, service handbooks and manuals provide a worthwhile alternative.

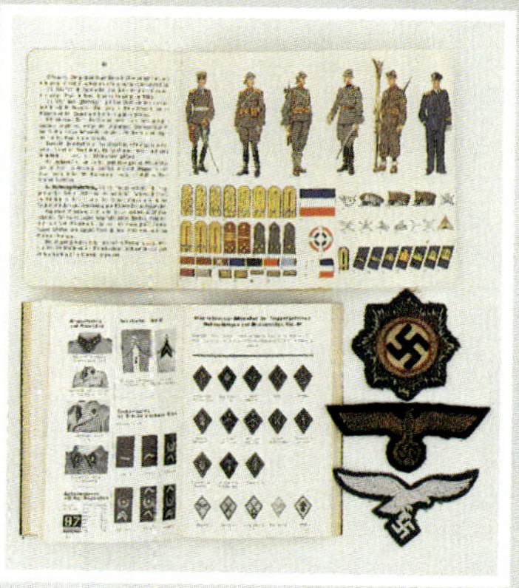

Uniforms and rank insignia of Yugoslav Army, 1941: German booklet issued to invasion forces. Pocket book issued for Swiss Army reservists.

13 A British Army Artillery Service Manual dating from before World War I. Many British artillery manuals published before 1914 are splendidly produced with fine colour art-work and are now highly prized for the illustrations alone. They are consequently rare, and fetch high prices when complete and in good condition.

14 Like (13), this example shows a fold-out from a vintage British manual on the 3·7in Mk 2 anti-aircraft gun. The wealth of detail contained in such publications is invaluable to the researcher and modeller alike.

15 *Home Guard*, a training manual for the British volunteer home defence force of World War II. When the Home Guard was formed in 1940, no provision was made for training material, and commercial publishers produced manuals for purchase by units or individuals. Many were of dubious value, but others, especially those written like this one by the novelist John Brophy. were excellent. They are not too difficult to find today.

16 A training manual for Soviet Artillery, published in 1982. The Soviet Union is so vast that the dissemination of training material through the Red Army has to be by central publication houses. By some quirk of the Soviet distribution system, this general artillery training manual can be purchased in the specialist Soviet bookshops found in the UK and Europe.

1 *SS-Soldbuch:* this document was issued to every recruit to the Waffen SS, the semi-autonomous combat units of the Schutzstaffel (SS) that operated with the regular German Army in 1939-45. A record book and identity document, it lists on the pages shown such details as the name (now indecipherable), service numbers, blood group and gas mask size. The Waffen SS had a strength of around 600,000 by December 1944, and since many old soldiers keep their papers such books are not uncommon-although they are not easy to obtain.

2 Canadian Army Paybook, World War II. As in most armies, the Canadian soldier was issued with a book in which all pay, allowances and deductions were recorded.

3 *Soldiers', Sailors' and Airmen's Prayer-Book,* issued for the use of Canadian servicemen in World War II at the recreational centres run by the Knights of Columbus, a fraternal charity organisation. "There are no atheists in foxholes" was a famous saying of World War II, and, although their behaviour sometimes suggested otherwise, many soldiers of all nationalities carried with them testaments or prayer books.

German Air Force briefcase, stamped "Luftwaffe Property"; with a 55-page German High Command Intelligence Report, dated August-October 1940, detailing bombing targets in the London area. Also shown: side cap of an NCO, Luftwaffe Engineer, and a navigation instrument.

4 Home-made cigarette lighter, World War II. Wartime restrictions in the combatant countries meant that luxuries like lighters were often unobtainable; so many servicemen made their own.

5 RAF Demobilisation Papers, officially dated November 1946. "Demob" was the dream of every "hostilities only" serviceman, so it is not surprising that many "old sweats" have preserved their Release Authorisation papers!

6 *Training Pocket Book* issued to the Royal Engineers, World War I.

7 Work Permit issued to a German civilian in 1940. Among the great number of official papers that every citizen of Germany and the occupied countries had to be ready to produce at any time was a work permit stating exactly what its

possessor did in the way of war work and where he was registered to carry on his work. As may be seen, such documents were replete with the official handstamps and signatures so beloved by the German bureaucracy. Very few seem to have survived.

8 Official publicity photograph of a World War II aircrew member of the Royal Air Force.

9 *Deutscher Volkssturm Soldbuch:* the personal record book and identity document (see also 1) of the German "People's Militia" set up by Hitler's order of 25 September 1944. A measure of desperation as Germany's war situation deteriorated, this conscript force was largely made up of those too old, too young, or otherwise unfit for regular service.

10 British *Soldier's Service and Paybook;* one of the most common of all service souvenirs, since this formal record of service was kept for every man and presented to him on demobilisation.

11 *A Short Guide to Great Britain,* published by the US War and Navy Department for issue to American servicemen posted to the UK during World War II. Most servicemen posted to foreign countries in that period were leaving their homeland for the first time, so such guides to the "natives" and their customs were necessary. Although thousands of these booklets were issued, they are not often found today.

12 A similar educational publication of World War I, *What a British Soldier wants to say in French.* Since this commercially-published booklet of 1914 was edited by a clergyman, it doubtless did not have the scope and application that many soldiers might have wished!

13 Cap badge of the Royal Australian Air Force, World War II. Note the dark, non-polishable, finish.

14 Finely-crafted, commercially-produced model of an RAF pilot, dating from around 1942. A fairly rare item, since the resources for such luxury products were in short supply during the war years.

15 Royal Air Force cap badge, with the "King's Crown" as worn in both World Wars.

16 British aircrew survival whistle, World War II. These were intended to attract attention to airmen who had ditched at sea.

17 Pilot's Wings, Royal Australian Air Force, World War II.

18 Pilot's Wings, Royal Air Force, World War II.

19 *Merchant Navy Clothing Book,* World War II. War rationing limited everyone in Britain to a set amount of new clothing, but the sailors of the Merchant Navy often needed more than the civilian norm to carry out their arduous duties and were issued with special allowance books.

20 *Pilot's and Flight Engineer's Notes,* an official British publication for aircrews operating the US-built Liberator (B-24) bomber. These books were issued for many types of aircraft and were usually in the "restricted" classification. Original copies are now quite rare—but many have been commercially reprinted for aircraft enthusiasts.

Military Postal Material of World Wars I and II

1

2 3 4 5 6 7

1 Two postcards from a set of eight, produced from official war photographs and sold as the *"Daily Mail" Official War Postcards, Series 21,* price 6d (2½p), in the UK after World War I. The cards show the devastation of the Somme battlefield in 1916.

2 A warning of the peril to come is given on this British local mail envelope, postmarked 17 May 1939 – some four months before the outbreak of World War II. The franking "AIR RAID PRECAUTIONS" was one of the methods used by the authorities to make the public aware of the danger of air attack and the need for Civil Defence measures.

3 All mail sent by servicemen during World War II (and in World War I) was handed in for censorship before being forwarded by the Field Post Office. The censor checked thoroughly to see that no man included information – notably the location of his unit – that might be of use to an enemy.

4 A sad task for the Army Post Office, New Zealand Forces, Middle East. The letter posted to Private Condon from the UK on 12 February 1942 arrived after he was killed in action: the laconic, pencilled "K.I.A." is reinforced by an official stamp for the letter's return to the sender.

Letters from home are vital to the morale of servicemen; thus, this traditional canvas-and-leather US Mail Bag, dated 1944, probably made an important contribution to the war effort.

5 Patriotic German postcard from a popular series of 1916: heroic artillerymen – one armed with an improvised pike – fight off the Allied hordes!

6 (Left) Posted just after the war, this piece of US Forces Mail was not subjected to censorship. The postmark shows that it was handed in at APO 462: the Army Post Office number denotes the region; in this case, Minneapolis, Minn. (Right) A small mystery: the claret 12-pfennig stamp was issued to commemorate Adolf Hitler's 53rd birthday, 13 April 1942 – but the Berlin postmark is 20 April 1941.

7 This rubber stamp – its impression shown to the right on a modern British envelope – is reputed to have been in use at the Afrika Korps

8 9 10 11 12 13 14

HQ of Field Marshal Erwin Rommel, 1940-43.

8 Air Mail Letter designed for the use of those writing from the UK to British prisoners-of-war (PoWs) in Germany and sold at Post Offices in Britain in World War II. It is a triple-folded sheet with a tuck-in flap to facilitate censorship in both countries.

9 First Day Cover of a US postage stamp issued to honour the US armed forces just after World War II. The stamp shows the Victory Parade in the Champs Elysées, commemorating the liberation of Paris by the Allies, 26 August 1944. The inset photographs show the capture of the Remagen Bridge over the Rhine by the US 1st Army on 7 March 1945.

10 High-value British stamps bearing the head of King George VI, issued during World War II.

11 US Air Mail letter, postmarked 23 October 1943 at AP 702 (Seattle, Wash.) and bearing the US Army censorship stamp. The 3c postage stamps feature the "V for Victory" sign, formed by the US eagle, with the slogan "Win the War".

12 These books of stamps issued by the British General Post Office date from 1952-53; postal rates were increased in the immediate post-war period.

13 Use of the double-fold Forces Letter by those writing to servicemen was encouraged by the British authorities, to save weight in aircraft and ships.

14 Postcards from two World Wars; (left) As their badges show, these soldiers belong to the British Army's Machine Gun Corps, which dates the photograph to the Corps' period of existence – 1916-19; (right) "Our Gracie" – the British popular singer Gracie Fields, whose great hit of the pre-war period, "Wish Me Luck As You Wave Me Goodbye" remained a favourite – partly through its apt title – in Britain throughout World War II.

(Above) A more detailed view of a Prisoner-of-War letter form as shown at (8) above; note the warning on the flap. (Below) This hand-drawn and -coloured birthday card was sent by a British PoW in Italy to his wife – the mother of the collector who now possesses it – in September 1941.

1 *Tripoli Times*, 15 June 1943; an extremely well-preserved specimen of World War II newsprint. This was an unusual publication in that it was both a service newspaper for Allied troops and a news sheet for local civilians, priced 2 lire.

2 *British Morning News*, 5 January 1946. Moving across Europe in 1945, the victorious Allies often found all local news media destroyed; they set up their own newspapers, like this Vienna-based publication, in order to maintain some kind of popular information service.

3 "The New Dance – the European Peace Walk". This odd postcard, marked as being found in Menton in southern France, may be based on a newspaper cartoon or on a café mural. It appears to date from the time of the Munich Agreement, September 1938, since it shows the leaders of the four powers concerned – Chamberlain (Britain), Hitler (Germany), Mussolini (Italy) and Daladier (France) – dancing a jig.

4 *Versprechen! (Promise!)*, one of many small booklets published by the German NSDAP ("Nazi" Party) to convey the party message to both soldiers and civilians. These were issued free and, although they are not rare, such is the appetite of collectors for Third Reich items that they are now difficult to obtain.

5 *The Crusader*, 26 December 1943: a free newspaper-magazine produced for the men of the British Eighth Army, first in Alexandria and then, following the invasion of Sicily and Italy in July-September 1943, in Tripoli. Few copies survive.

6 *Eighth Army News*, 24 December 1943. Another news sheet produced for British invasion forces in Italy; and, like (5), now rare.

7 British official photograph taken during an RAF leaflet raid: an airman dispenses propaganda leaflets (see 8, 12, 13, 15) in bundles that will break up in the slipstream of the aircraft.

8 British propaganda leaflet, typical of the many types dropped over Germany and the occupied countries. This one is aimed at workers in the Ruhr, Germany's great industrial area, and calls on them to revolt against the Nazis or to carry out sabotage. Such leaflets now represent a highly specialised branch of militaria collecting.

9 *Union Jack*, 21 July 1944: yet another locally-produced newspaper for British forces in Italy. One of the best of its type, it was very popular and surviving copies are relatively common – although copies in good condition are rare.

10 Political List issued by the NSDAP. The Nazi Party was prolific in its output of publications: lists like this were produced in many forms, including the names of Party officials and members, proscribed persons, edicts and announcements. However, since they were often printed on poor-quality paper, specimens in good condition are now eagerly sought after.

11 *Eighth Army News*, 23 July 1945. The war in Europe was now over and troop levels in Italy, where

this issue was published, were falling. Thus print runs were smaller and papers of this period are uncommon.

13 German propaganda leaflet dropped on Allied invasion troops in France, late 1944. It argues that the war is being waged solely to benefit the Soviet Union and that the Allies have walked into a trap: in view of the Luftwaffe's shortage of aircraft at the time, a fairly wasteful and unconvincing piece of work.

14 *The Prisoner of War*, a journal intended not, as its title suggests, for British PoWs, but to give news and encouragement to their families at home. Published in London by the Red Cross and St John's Ambulance amalgamated organisation, this journal was often carefully saved by those who received it and is still extant in some numbers for the collector to find.

15 Allied propaganda leaflet dropped over Sicily and Italy at the time of the invasion of the former territory, July 1943. Well-printed in colour, it should have caught the eyes of the Italian people, to whom the message was simple: kick out the Germans before the Allied invasion of the Italian mainland.

Service cap of the kind worn by British War Correspondents, both male and female, during World War II. Apart from the distinctive cap badge, it is identical to the service cap worn by British officers. War Correspondents also wore the same uniform as officers, but with an identifying flash on the shoulder.

1-4 Currency notes brought home from the Far East by soldiers or former prisoners-of-war as souvenirs. (1) Chinese note; (2) Ten cent note issued by the British administration of Hong Kong, the British crown colony that fell to the Japanese in December 1941; (3) Japanese note; (4) Japanese occupation currency note, used in the territories under their control and generally referred to by Allied troops as "banana money".

5 German banknote for 50 million marks, dating from the great inflation of the 1920s, the ravages of which did much to increase support for Hitler's "Nazi" (*Nationalsozialistische Deutsche Arbeiterpartei* (NSDAP) = National Socialist German Workers' Party) faction. (See also 9.)

6-8 Paper currency issued for the use of Allied troops during and after the liberation of France and occupation of Germany, 1944-45.

9 To counter the ravages of the post-World War I inflation, the German authorities issued currency based on land values, called *Rentenmark*, from 1923.

10 British Army *Prayer Book* printed in 1918 for the Chaplain General. It contains a selection of prayers and hymns suitable for parades.

11 US Army manual *TM 30-306, German Language Guide*, issued in 1944-45.

12 *Passierschein:* Safe Conduct Pass, "valid for one or several bearers", air-dropped behind the German lines after the Allied "breakout" from the Normandy beachheads in August 1944. Signed by the Allied Supreme Commander General Dwight D. Eisenhower, USA, the object of the pass was to encourage Axis troops to desert the Reich.

13 *Air Raid Precautions Handbook No 2,* printed and published by His Majesty's Stationery Office (HMSO) and issued by the Home Office in 1935 – some four years before the threat of air attack on Britain became a reality. In these plates, a member of the St John's Ambulance organisation models the Special Service Respirator.

14 The photograph holder is a typical example of the British "utility" artifacts of World War II. The photograph shows an RAF aircrew man of the early World War II or immediate pre-war period in a Sidcot flying suit and "zip-eared" flying helmet.

15 Souvenir Programme of the great parade staged in The Mall, London, on 8 June 1946 to commemorate the first anniversary of the defeat of Germany (although "V-E Day", Victory-in-Europe Day, was in fact 8 May 1945).

16 This photograph, dating from around 1941, shows a group of candidates for the Royal Navy's "Y" Scheme at HMS *King Alfred,* a naval shore establishment at Hove, Sussex. Candidates who were thought to have leadership potential were recruited from schools or civilian jobs (note the absence of medal or campaign ribbons) and given special training. The gaiters worn suggest that this was a gunnery course. Military photographs are a rewarding – and sometimes relatively cheap – field

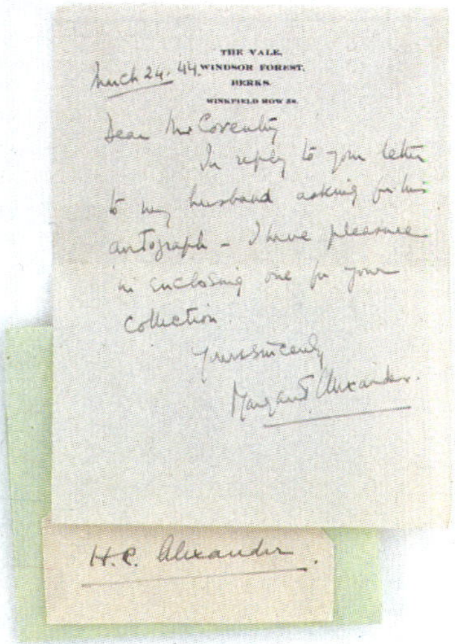

for the collector; the research needed properly to identify them, as in this case, can be both fascinating and time-consuming!

17 Sheet music of a popular song of World War II, published in 1943. The title comes from a phrase used by US aircrew when returning to base with a damaged aircraft.

18-20 Victory Medal, World War I, awarded posthumously to Sergeant G. Bishop, Royal Field Artillery. The medal, with its citation (19) dated 2 September 1921, was sent by registered post (20) to his next-of-kin. The interest of a common medal is obviously greatly increased by possession of such poignant documentation.

21-22 Clothes rationing was introduced into wartime Britain on 1 June 1941. Every person with a ration book was issued with 66 coupons, as shown here, annually. Typical values included 16 for a man's overcoat and 8 for a shirt.

23 *Le Courrier de L'Air*, a news sheet printed in French in London and dropped by the RAF over occupied France. Newspapers in the occupied countries printed only what the German authorities would allow – and the headline "Ruhr torn apart by bombs" in this issue of 24 June 1943 would not have come into that category! The subject populations were forbidden to listen to the broadcasts of the British Broadcasting Corporation – but the transmission times and wavelengths were given here.

24 The present-day collector will wish that these prices were still current! This price list of Nazi regalia was issued in 1930, some three years before Hitler came to power. Here, his supporters are offered Party pins and pendants; SA (*Sturmabteilung;* "Brownshirt") uniforms; and camping gear.

25 *Illustrated,* issue for the week ending 30 December 1939. This popular weekly, somewhat similar in format to its famous contemporaries *Picture Post,* in Britain, and *Life,* in the USA, was published by Odhams Press, London. The cover picture of this issue refers to the main photo-story: "Tommies' Christmas", with the British Expeditionary Force in France.

The autograph of Field Marshal Earl Alexander of Tunis, sent by Lady Alexander to a collector during World War II.

1-2 Contemporary US Army Recruiting Leaflets for (1) 2nd Armored Division, nicknamed "Hell on Wheels"; and (2) 1st Infantry Division, nicknamed "The Big Red One". Both bear reproductions of the units' shoulder sleeve insignia. Such material is cheap, colourful and relatively simple to collect; leaflets and similar publicity material can often be obtained for no more than the cost and effort of a postal request to the service or unit concerned. See also (6).

3 *Yank* magazine; Continental Edition, dated 14 January 1945. Perhaps the liveliest and most interesting, and certainly the widest-circulating services publication of World War II, *Yank* was originated by the US Army's Special Service Division in April 1942 and was published until mid-1945. Its head office was in New York, but no fewer than fifteen editions, covering all theatres of war, were printed bi-weekly worldwide. The magazine was intended for and written by the NCOs and Enlisted Men of the US Army; its material came from its own soldier-journalists and from voluntary correspondents. In standard picture magazine format, it included war news, home news and sport—and, of course, a full-page "pin-up" photograph—and, able to speak relatively freely, constituted an important and influential lobby for the welfare of the ordinary "G.I." A copy of its contemporary, the US Army daily newspaper *Stars and Stripes*, is shown on *pages 176-177*. The

Of interest both to specialist match-box and -book collectors and to the militaria collector: a match-book of World War I; an item donated to US servicemen by the American Red Cross.

equivalent publication for German servicemen of World War II, *Signal*, equally professionally produced, but with a far higher content of straight propaganda, is shown on *pages 162-163*.

3-5 Propaganda leaflets printed by the South Vietnamese authorities as part of the *Chieu Hoi* ("Open Arms") operation, which was aimed at persuading the communist Viet Cong carrying on a guerrilla war in South Vietnam to surrender with their weapons to the Free World forces. These leaflets, and those shown at (7) and (8), date from around 1969.

6 A further US Army recruiting item: a car bumper sticker for the 1st Cavalry Division, bearing the unit's shoulder sleeve insignia and a cavalry officer in the famous "Kitchener needs *You!*" pose of the British recruiting poster of World War I, in 19th-century uniform.

7-8 Further propaganda leaflets produced on behalf of the Government of South Vietnam as part of the *Chieu Hoi* programme. (7), bearing the flags of the Free World forces in alliance with the Republic of South Vietnam, is a safe-conduct pass of the kind air-dropped in huge quantities on areas where the Viet Cong were suspected to be present in any strength.

9 US Civilian Ration Book, World War II; these were issued but, unlike Britain where food rationing persisted into the 1950s, were not brought into general use.

10 US Army *Field Manual FM21-150*, "Unarmed Defense for the American Soldier", dated July 1942: a basic pictorial guide to self-defence against armed and unarmed opponents. The original owner of the manual has scrawled across its cover the comment: "Army Tricks!"

11 Any wartime newspaper will be of interest to the collector of printed material—but most to be prized are those covering the major events of the conflict. This special ("Extra") edition of the *Moose Jaw Times-Herald* of Saskatchewan, Canada, is date-lined 6 June 1944 and carries the news of "D-Day", the invasion of the European continent by Allied Forces. As explained at length in the Introduction to this section (see *pages 162-163*), wartime newsprint is often of inferior quality and, like the newspaper shown here, will become discoloured and crack if it is not kept dry and stored, preferably in a plastic wallet, away from extremes of temperature.

12 As in the case of (11), this wartime edition of *Time* magazine, dated 30 October 1944, is of particular value to the militaria collector in that its cover subject is General Douglas MacArthur, Allied Supreme Commander in the Southwest Pacific Area, who, at the time of publication, was fulfilling his pledge of March 1942—"I shall return"—by undertaking the liberation of the Philippines from Japanese occupation. A full-size edition of *Time* is shown here: a special half-size edition, on lightweight paper and carrying no advertising, was produced for servicemen overseas.

1 Copy of US forces' newspaper *Stars and Stripes,* dated 3 May 1945, announcing the surrender of German Forces in Italy.

2 British Government publication *Meet the US Army.* Published for public sale in 1943 by the Ministry of Information as an introduction to the American forces who were then arriving in Britain in ever-increasing numbers preparatory to the invasion of Europe. Well illustrated with pictures of US soldiers, their weapons and transport, and the main types of US aircraft, the booklet also has an appendix with a basic list of American words and their differing British meanings.

3 US Army writing set: this pack of writing paper and envelopes was not an issue item but was sold in PX (Post Exchange) Stores at US camps and bases.

4 US Army manual *Company Duties — a Checklist,* published by the *Infantry Journal* magazine. First published in 1941, this booklet was intended for all NCOs and officers up to the rank of captain. A basic outline of their duties, it was intended as an *aide-mémoire* for the newly-promoted soldier.

5 US *New Testament* with steel plate front cover. These were made for sale to the wives and sweethearts of servicemen who were about to go overseas. Although the steel is not thick enough to stop a rifle bullet, soldiers possibly derived psychological benefit from carrying talismans like these, presented to them by their loved ones at home, in their combat clothing.

6 US Army *Driver Manual,* published by the US Government for trainee military drivers.

7 US Army Prophylactic Kit, intended for the prevention of venereal disease and issued to all overseas troops when they were granted leave passes.

8 US Army *Field Manual FM 23-35,* on the Automatic Pistol Calibre ·45 M1911 and M1911A1. This rare manual, published as late as 1940, was prepared under direction of the Chief of Cavalry and chapters deal with firing the pistol while on horseback, as well as methods of accustoming horses to the sound of gunfire at close ranges. Well illustrated with photographs of mounted cavalrymen, the book also has chapters on marksmanship and basic repair and maintenance of the pistol.

9 British manual on the Browning ·300-calibre water-cooled machine gun, published by Gale and Polden early in World War II and intended for use by Home Guard units.

10 US Army *Field Manual FM 23-40* on the Thompson submachine gun, Calibre ·45, Model M1928A1. Dated 31 December 1941 and published by the Chief of Cavalry, this was a basic instruction manual for recruits to the Cavalry or the newly-created Armoured Forces, with detailed instructions on how to load, strip and fire the "tommy gun". It is interesting to note that none of the recommended firing positions includes firing from the hip — despite the example of Hollywood gangster movies in which this method is invariable.

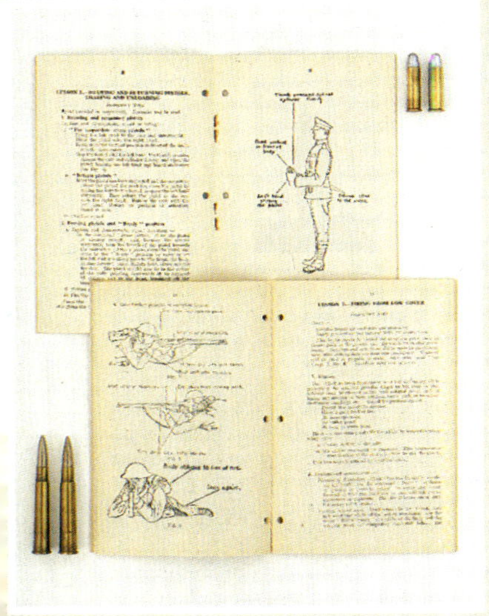

11 Identity Card for Military Assistance Command Vietnam Studies and Observations Group. This item, although a reproduction of the originals used in Vietnam by Special Forces and CIA agents involved in intelligence operations against suspected Viet Cong and sympathisers, is collectable as a display piece in view of the extreme rarity of the genuine article.

12 US Army *Pocket Vocabulary* of Malay, Pidgin English and Japanese phrases, published in September 1942 for issue to troops in the Southwest Pacific Area (Solomon Islands, New Guinea and the Netherlands East Indies). The section dealing with Pidgin English phrases is particularly interesting: the Pidgin phrase for "Don't move or I'll shoot" is revealed as being "You-fella stand fast. You no can walkabout. Suppose you-fella you walkabout me killim you long musket!"

13 US Army issue *New Testament, Protestant Version.*

14 US Army *Field Manual FM 23-7* on the US Carbine Calibre ·30 M1 and M1A1, published in April 1944 and intended for recruit training on the M1 Carbine. Well-illustrated chapters deal with the loading, firing and stripping of the weapon, and a section is devoted to the M1A1 version, which had a folding stock and was issued to paratroops.

15 US Army *Pocket Guide to Vietnam.* Issued to troops newly posted to South Vietnam, this booklet of the 1960s gives a short outline of the history of the country and the customs of its inhabitants. Of particular interest to the collector in other fields than printed material is the booklet's basic guide to South Vietnamese Army, Navy and Air Force uniforms and insignia. There is also a short vocabulary of commonly-used words and phrases in Vietnamese.

16 US Army manual of World War II on the steel helmet: *It's Your Head – Keep It!* Issued to new recruits, this was a basic set of instructions on how to adjust and wear the protective steel helmet.

British weapon training manuals of World War II vintage for the Webley ·38in pistol and No 4 rifle. As well as being collectable in their own right, manuals give a useful insight into the tactical doctrines of their publication period.

1 *Lustige Blätter* ("Funny Pages"), War Number 32, published 1915. This famous satirical journal was not unlike the British weekly *Punch*, which also published many patriotic cartoons during World War I. The left-hand page shows *"Unser Herkules"* ("Our Hercules"), Field Marshal von Hindenburg, "persuading" a Russian officer to leave East Prussia, where Samsonov's Russian armies had been crushed by German forces led by Hindenburg and Ludendorff at Tannenberg, August 1914.

2 *Der graue Ritter* ("The Grey Knight"); a popular novel of the Western Front by Karl Rosner, published in the 1920s. Since many Germans were embittered by the harsh terms of the Versailles Peace Treaty, novels that glorified the bravery of German soldiers ("grey knights" from their "field-grey" uniforms) were extremely popular. See also (8).

3 German Army weapon instruction manual of c1910; the illustrations show the breech and bolt mechanism of the Gew 98 rifle.

The impressive presentation portfolio and citation for the award of the Knight's Cross of the Iron Cross to Colonel Erwin Menny, commander of the 15th Rifle Brigade, is dated 26 December 1941 and bears the signature of Adolf Hitler. Many men received only provisional citations—issue of the documents seen here was to be made after Germany's victory—but Colonel Menny was among the lucky ones.

4-5 Collectable in themselves—but also invaluable aids for the research entailed in many branches of militaria collecting: German Army Rank-Lists for (4) the Royal Army of Saxony, 1914, and (5) the Prussian Army, 1914. These were published annually and listed all regiments and their officers; the Prussian List was established under Frederick the Great in the 18th Century and continued to appear until 1918. With the aid of such lists it is possible to trace the entire course of an officer's career.

6 Prisoner-of-War escape stories have always been a flourishing branch of popular military literature. *Unsere Flucht aus französischer Kriegsgefangenschaft* ("Our Escape from French War-Captivity"), the "personal experiences" of Gustav Kreipe and August Siegmann, was published in Berlin soon after World War I; see also (2) (13) (14) (15).

7 *Der Drahtverhau* ("The Wire Entanglement"); a trench newspaper published by the 3rd Company, Bavarian Reserve Infantry Regiment No 1, in 1916. The high quality of production of this journal suggests that it was the work of soldiers in a fairly quiet sector of the line!

8 *Die Grosse Frühlingsschlacht 1918* ("The Great Battle of Spring 1918"); a popular account of the massive and unsuccessful last-ditch onslaught by the German armies on the Western Front in 1918. By Karl Rosner, author of the novel at (2), this was another of the many publications aimed at restoring German national pride in the 1920s.

9 Embossed paper seals, replacing the former impression of a signet in wax, used on official Prussian military correspondence from about 1900.

10 Bavarian Army manual of the pre-World War I period, showing rank and trade insignia. Since various German state armies are mentioned on this spread and elsewhere in this book, it is worth noting that the Bavarian Army, for example, was both the army of the Bavarian state and, from 1871, part of the Imperial German Army under the wartime leadership of the Kaiser.

11 *Unsere Bayern im Felde* ("Our Bavarians in the Field of Battle"); a part-work (fifty weekly issues) reporting the deeds of Bavarian units and soldiers in the campaigns of 1914 and 1915. Such journals—and their Allied equivalents—circulated widely among civilians whose patriotism was kept at a high pitch by this and other propaganda.

12 German Army *Dienstaltersliste* (Service and Seniority List) of 1914; listing all officers with dates of birth, commission and promotions.

13-15 Three further examples of the popular war literature published in Germany soon after World War I: (13) *Zeppeline über England*, an account of the bombing raids made by German airships; (14) *Hans Hellers Höllenfahrt* ("Hans Heller's Journey to Hell"); presumably an "anti-war" tract; (15) *Stürme und Luftsiege* ("Storms and Air Victories") by Captain Ritter von Tutschek, Royal Bavarian Army, a pilot of World War I.

Military Miscellanea

Joe Lyndhurst

Apart from the German military souvenirs shown on *pages 184-185*, the wide and varied selection of miscellaneous militaria shown on the following pages was assembled by one collector, the present writer, as a part of the Warnham War Museum. This selection will, I hope, serve to show just how many kinds of military collectables exist outside the artifacts classified under the various special headings in the earlier parts of this book. A brief summary of my own experience as a collector may also illustrate the way in which the collecting process may develop and the collection grow.

The miscellanea shown on these pages range from severely practical items like the alloy mine probe (*pages 204-205*) of the kind still in use in the aftermath of Britain's reconquest of the Falklands in 1982, through the personal, like the Japanese soldier's "comfort fund" fan (*page 195*) of World War II, to the frivolous, like the rather improper effigy of Adolf Hitler made by an Italian prisoner-of-war (*page 183*). However, all the items have one thing in common: whether impressive or trivial, they help to reflect just what life was like at the times of the major conflicts of this century and thus play an important part in telling the full story of war. To my mind and, I am sure, to those of most collectors, this makes them eminently "collectable".

COLLECTOR'S PROGRESS

My own collection of miscellanea began as a result of my activities at Warnham. My original intention was to form a Museum of Military Transport; to illustrate the development of mechanised transport and show how the lessons of World War I were put to good use in World War II. I soon realised that a vehicle, however fascinating in itself, is still just a vehicle and needs certain embellishments to give it life and character. I found that it was necessary to introduce a "human touch" by adding personal equipment to the display. And thus I entered a field of tremendous interest in which there was the ever-present excitement of discoveries to be made.

Properly to display a platoon truck of World War II, I needed a vehicle tool kit, a rifle to put in the rifle rack, a kit bag or two, a box of rations, a camouflage net, webbing, gas-masks and an entrenching tool; all these on the vehicle itself, with appropriate uniforms, personal gear like shaving kits and weapon and equipment cleaning kits, and documents—a military driving permit; the vehicle's lubrication chart —in a cabinet nearby. To show a Special Air Service (SAS) jeep dating from the unit's formation in the Western Desert in World War II, I needed such impedimentia as German "Jerry cans", a sun compass, a small cooker, towing chains, sand mats (for extricating the vehicle from soft sand), water bottles and ammunition boxes.

Of course, most collectors will have neither the space nor the resources to form a collection of vehicles, but very worthwhile collections can be made of smaller items of the kind I have mentioned, which have been roughly grouped thematically on the following pages. The new collector might consider, for example, assembling an array of compasses and maps (*pages 200-201*), torches and signalling lamps (*pages 202-203*), or military issue tools (*pages 204-205*). The point is, that in militaria collecting one item seems inevitably to lead to another: acquire a jacket, and you will want a hat to go with it, and a badge to put on the hat, and spare buttons for the jacket, and a "button-stick", and a webbing belt . . . and so on. A collection *grows!*

BUYING AND RESEARCHING

In the late 1950s and early 1960s, when I began collecting, I was able to find many military items of the kind I needed for my displays in junk shops, which were probably a better source for collectors then than they are now, when, both in Europe and the USA, they tend to call themselves "flea markets" and to raise their prices. However, it is still possible to find bargain collectables at jumble sales, garage sales, war surplus

Far left: *The US Army's BC-611-c radio, the famous "walkie-talkie" (see (11),* page 203*) had a great influence on small unit tactics. It is seen here in use by a US Marine on Iwo Jima, Pacific, March 1945.*

Left: *A US soldier eats a hurried meal from a GI mess-tin (see* page 197*) during the "Battle of the Bulge", January 1945.*

Right: *Binoculars and map-board (see* pages 200-201*) are carried by a scout-plane pilot aboard the heavy cruiser USS* Chester *(CA 27), in the Pacific, 1942.*

Below: *Posed in a training area in France, in 1918, this dramatic representation was intended to remind US soldiers of the dangers of gas attack and of the need for respirators (see* pages 190-91).

shops (handy for webbing and small items of uniform) and—a fairly recent innovation, and a boon to collectors—militaria collectors' fairs.

As material accumulates it will become obvious, as it soon did to me, that reference sources are needed for the proper identification of articles of uniform and equipment. In World War II, for example, uniforms and equipment differed not only between countries and services but often within them, according to campaigns and theatres of war. The British Army had several styles of battledress; New Zealand soldiers' battledress was of slightly "greener" khaki than that made in the UK; Canadian battledress differed again. Different types of clothing were issued for use in different theatres of war, from the Arctic to the Tropics, from Murmansk to the South Pacific. The best guide to the subtle variations in uniform and equipment is the study of contemporary photographs. Reference works that are likely to be of use to the new collector are listed on *pages 206-207,* but in the 1960s, when there were fewer military reference books available than there now are, I found that a very useful source was the magazines published for modellers, which contain in-depth reference on such arcana as vehicle markings and the minute details of uniform and equipment needed for painting individual figures.

A WEALTH OF MATERIAL

As the following pages will show, a wealth of material is available for the general collector. A few specific remarks about these items may help the newcomer to the field.

Of all the items shown, the most difficult to find were the perishables. Foodstuffs (*pages 196-197*) are so important in wartime that they are obviously of interest to the collector, but most wartime rations now only survive in tins. Dried egg and dried milk, packed in the USA, Canada, Australia, South Africa and elsewhere, may be found, as well as the triumph of the food processors' art: the carefully-balanced field rations packed for servicemen, which may include such interesting items as self-heating cans. An interesting collection could be made of mess-tins and eating utensils, with such additions as the German soldier's coffee-grinder (*page 185*) and the Jewish Brigade's "Passover bag" (*page 192*). That other necessity of the fighting man, tobacco, probably survives in greater bulk in the form of cigarettes, even dating back to World War I. Souvenir cigarette cases, especially if engraved with individuals' names and units, make an attractive addition to a collection, as do cigarette lighters made from empty cartridge cases or matchbox holders crafted out of scrap metal (which are shown on *pages 182-183*).

Servicemen of all nationalities, although generally not well paid, have always sought to bring home souvenirs of their travels and campaigns. A wide selection of these is shown, ranging from the elaborately-decorated porcelain drinking mugs (steins) and pipes favoured by German reservists of Imperial times (*pages 184-185*) to a US-made "Zippo" cigarette lighter (*page 192*) carried by a British serviceman in the Falklands campaign of 1982. Perhaps the most attractive and "authentic" examples of servicemen's souvenirs fall into the category of what has become known as "Trench Art", from the handicrafts of soldiers in the trenches of the Western Front during World War I. As may be seen from the fine examples of British and German trench art shown on *pages 182-185,* many of these artifacts were made from the most readily available materials, cartridge cases and shell fragments; the scrap brass and copper is often finely crafted, probably with the help of unit workshops or blacksmiths. Embroidery,once a speciality of the Royal Navy, samplers worked by soldiers and bearing unit badges and colours, and wood-carvings are also to be found.

Cheaper to acquire, and most attractive, are souvenir postcards and handkerchiefs of the kind shown on *pages 186-189:* of greatest interest to militaria collectors will be those worked in silk with unit badges. Mostly dating from World War I, these were, of course, commercially produced—but such articles, as distinct from those of military issue, are not to be scorned by the collector. Such commercial items as the "Victory teapot" of World War II and the cushion-cover with sewn-on silk unit badges of the kind given away with cigarettes during World War I (both shown on *pages 192-193*) would grace any collector's shelves.

A reminder of the darker side of war is given by the collection of gas-masks shown on *pages 190-191.* It is fascinating to see how different countries planned to protect their servicemen and civilians in the event of gas attack, a nightmare of World War I that did not, happily, again become reality during World War II. Gas-masks for civilians with special requirements, like that for the Post Office Telephonist, are particularly interesting; while the gas-mask for a baby, virtually a pressurised suit, is a chilling reminder that war is no respecter of the innocent. The reality of war is also brought home by the display of medical items on *pages 198-199;* as an international organisation, the Red Cross, active during every conflict of this warlike century, would make a good field of specialisation for a collector of any nationality.

British "Trench Art" and Souvenirs, 1914-1945

1 Silk and wool pin-cushion bearing the badge of the Machine Gun Corps, made by a serving soldier as a souvenir to bring home from the Western Front during World War I.
2 Pincushion token of World War I, with the badge of the Royal Field Artillery; possibly made by an artilleryman while convalescing after a wound.
3 Sailors are traditionally producers of fine handicrafts: this Royal Navy pin-cushion is no exception.
4 Matchbox cover made from scrap brass by a British soldier serving in the Peronne area on the Western Front, 1918.
5 Steel matchbox cover incorporating the cap badge of the Gloucester-shire Regiment.
6 This brass matchbox cover has the upper part of a tunic button of the Royal Engineers soldered to its face.
7 Brass matchbox cover depicting the German Iron Cross, dated 1914.
8 Sheet copper was used for this matchbox cover. The date 1919 refers to Allied intervention against the Bolsheviks in Russia.
9 Commercially-made matchbox cover showing the German Army's characteristic *pickelhaube*, or spiked helmet.
10 This matchbox cover is of aluminium. Since DUNKIRK is inscribed on it, the metal may have been salvaged from the nearby World-War-I aircraft repair depot.
11 This brass matchbox cover incorporates part of the cap badge of The Royal Scots Greys (2nd Dragoons), a cavalry regiment.
12 This commercially-made post-war matchbox cover commemorates the battles for Hill 60 at Messines Ridge in 1917.
13 Brass matchbox cover made by a Royal Artillery gunner, c1914-18, with an officer's collar badge soldered to the face.
14 Pen made from a German brass cartridge case, c1914. The relief nib can be placed inside the case when not in use.
15 A very popular type of cigarette lighter—a large brass nut with an old penny soldered on either side—probably made in a Royal Electrical and Mechanical Engineers workshop, c1939-45.
16 German-made matchbox cover with what appears to be the owner's name and his home town (Görlitz) engraved along the edge; possibly the work of a prisoner of war.
17 An attractive cigarette lighter made from brass hexagon stock, obviously produced on a lathe in a workshop and probably a product of World War II.
18 Cigarette lighter made from a US ·50-calibre round, World War II.
19 Commercially-made petrol cigarette lighter of World War II: these cheap mass-produced items helped to counter a severe shortage of matches.
20 Souvenir toasting-fork, made from a British ·303in cartridge case of World War I, with a Royal Artillery badge—part of a small button—soldered on to its handle.
21 Letter opener made from a British ·303in cartridge case, 1914-18; another souvenir made by a serving soldier.

22 The copper driving band of a World-War-I shell forms the handle of this brass letter opener.

23 Commercially-made lady's powder compact in enamel showing the flags of the Allies, World War II.

24 Cribbage board made from the butt of a British Lee Enfield No 4 Mk 1 rifle, World War II.

25 "Trench art" at its best: a German shell case transformed to the likeness of a dress cap of the Grenadier Guards, 1914-18.

The tin box to contain one day's rations for three men is of the type carried in British armoured fighting vehicles, 1939-45. The round mess tins are dated 1941, but are of the pattern designed for cavalrymen to strap on their saddlebags. The final, vital, piece of equipment is a corkscrew bearing the War Department stamp.

26 A useful gift from the Front: a pipe-cleaning set in a German cartridge case.

27 Four ·303in bullets frame the arms of Albert, Belgium, on this World-War-I ashtray.

28 Favourite smoke of two World Wars: five Woodbine cigarettes in paper packet; price in 1939 – 2d (0.8p).

29 Cigarette case fashioned from scrap aluminium from a crashed aircraft by a Japanese soldier in Burma during World War II, and subseqently "liberated" by a British or Commonwealth soldier.

30 This aluminium cigarette case made for a British soldier stationed near a prisoner-of-war camp is typical of the handicrafts of Italian PoWs in the UK, 1939-45.

31 Packet of 12 German cigarettes of World War II.

32 Wartime model in 1/72 scale of a Spitfire, in wood, with lead propeller, solid perspex canopy and wheels of the type supplied in pre-1939 "Skybird" kits.

33-34 "Trench art" of the Royal Flying Corps during World War I: (33) brass biplane with a ·303in round forming the fuselage; (34) a French cartridge forms the fuselage of a SPAD biplane.

35 Humorous "toy" made by Italian PoWs in the UK during World War II: when the coffin lid is slid back, a strategically-placed piece of valve rubber springs erect!

36 This extremely attractive platter is made of folded and interlocked cigarette packets and is another Royal Navy product, World War II.

1 Trench art of World War I: the Iron Cross features on this vase carved from an ox-bone by a soldier.
2 German *Reservistenflasche* ("Reservist's Flask"). Bottles like this—of porcelain with sheet-zinc edging, on a lanyard in the black, white and red colours of the Imperial German Reich (Empire)—were made by many manufacturers between 1871 and 1914, for sale to conscripts as a memento of their service time.
3-5 Pipes were especially popular as souvenirs of service for German soldiers. A typical example is shown at (3): made for a reservist in the 3rd Baden Dragoon Regiment, stationed at Karlsruhe, 1893-96. The bowl is porcelain, with a metal lid in the shape of a *pickelhaube* (the German spiked helmet). The pipe at (4), a porcelain bowl with a silver base and rim, portrays the Emperors Wilhelm I, Friedrich III and Wilhelm II. At (5) is a meerschaum pipe with the bowl portraying Wilhelm I.
6-7 Letter-openers made from the copper driving-band of a shell by a German soldier of World War I.
8 This silver cigarette case contains an embroidered leather photograph frame and leather card case; the latter bearing an inscription identifying the items as a reward for marksmanship by Musketeer Populoh, 1903.
9 A *schnaps* (liquor) flask of decorated earthenware in the shape of an artillery shell; "42er" on the label refers to the flask's volume. A patriotic souvenir, commercially produced early in World War I.
10-11 Defused 8·5cm shells, with detonators, decorated by German soldiers, World War I. (10) shows battle scenes of 1916 campaigns; (11) shows scenes from the life of a field-artillery unit in 1914.
12 A rare and valuable plate of Meissen porcelain with cobalt blue decoration, produced to commemorate the great German victory over the Russians at Tannenberg in August 1914.
13 German sailor's handiwork of World War II: a model of an 11in (279mm) gun turret. Brass, on a wooden base, with a painted Iron Cross and the letters "KM" (*Kriegsmarine:* the German Navy).
14 Silver cigarette case presented to a German officer in World War I. The signatures of the subscribers are engraved on the exterior.
15 Small coffee grinder with detachable handle: many German soldiers carried these in their packs during World War I.
16 The bust of a "typical" Prussian militiaman forms this porcelain ashtray dating from *c*1900.
17 "German Empire Colonial Clock": a very rare souvenir of the German thrust towards an empire based on sea-power, dating from 1905. The inscriptions on the coloured metal case read: "The sun will not set on our Empire" and "Our future rests on the waves".
18 This copper snuff-box was made by the German General Nickelmann during his time as a PoW in Russia, World War II.
19-20 Iron letter-openers made from shell splinters; World War I.
21 Patriotic token: the hand of Field

Marshal von Hindenburg, with rank baton, hollow-cast in bronze by St Cauer, 1915.

22 Another symbol of patriotism for purchase by those at home, dating from 1914-16: an iron beaker with gilded medallions of Franz-Joseph I of Austria and Wilhelm II of Germany. The metal band, with facsimile of Hindenburg's signature beneath, bears the inscription, in German: "The iron cup, when full, is consecrated to the iron trio of the iron time".

23-26 Along with pipes, decorated *steins* (beer mugs) were favourite souvenirs of service in the Kaiser's army. These were all produced for sale to reservists. (23) belonged to a member of Ludwig-Wilhelm III's Baden Infantry Regiment No 111. Made of porcelain with a pewter lid, it is typical in that as it is emptied an inset scene in its translucent bottom becomes visible. (24) is from the 5th Baden Infantry Regiment No 113, Freiburg, 1900-02. (25) as the field gun and artillerymen on its lid suggest, is from an artillery unit: 2nd Baden Field Artillery Regiment No 30, Rastatt. (26) is an earthenware stein of the Baden Field Artillery Regiment No 14, Karlsruhe. Such articles fetch a very high price today: the collector should beware of modern copies masquerading as originals.

Bearing the cockade of Imperial Germany, these hats were worn by Schutztruppen (Colonial Troops) on home leave. (Above) ORs, German East Africa; (Below) Officer, Cameroun and Togo.

British "Sweetheart Cards" of World War I

These attractive postcards dating from the World War I period are now very popular collectors' items. They have no specific name: in Britain they are generally known as "Sweetheart Cards". Commercially produced, with the motif and/or message delicately embroidered in silk, they were sold in regimental canteens and similar outlets and bought by servicemen to send home to their loved ones. Since they were often carefully preserved as keepsakes, they are not uncommon today. As may be seen from this selection and from the further examples on *pages 188-189*, motifs varied from regimental crests and corps' insignia to sentimental messages and "souvenirs". As with the "Sweetheart Brooches" shown on *pages 56-57*, the colourful flags of the Allied nations often formed a part of the designs.

1 This card bears the rod-and-serpent insignia of the Royal Army Medical Corps (RAMC) – unkindly nicknamed "*Rob All My Comrades*" in reference to the supposed proclivities of medical orderlies. Comparison with (19) will show that the styles of these cards varied even within the same unit, and that the colours used were rarely the "official" unit colours.

2 "To my dear Sister": typical of the "family" messages, in the popular decorative style of the day. Like many of the cards shown here, this example was manufactured in France.

3 "Souvenir d'Arras". Arras was the scene of bloody fighting in both World Wars – and this card cannot have been of much comfort to the recipient, since it appears to show the Cathedral in flames as a result of World War I bombardment.

4 In attempting to reproduce accurately the Order of the Garter insignia and the hart on the ford

Italian souvenir handkerchiefs with patriotic designs. The smaller example, printed on silk and dating from World War I, shows Victory urging on an Italian soldier of the elite Bersaglieri and a militiaman. The larger example, dating from the 1930s, shows various aircraft of the Regia Aeronautica (Italian Air Force) with the arms of the House of Savoy in the centre and the Fascist party emblem in the corners.

that formed the badge of the Bedfordshire (later the Bedfordshire and Hertfordshire) Regiment, the needlewoman has made the wording somewhat indistinct.

5 "Forget Me Not": presumably no maiden of the 1914-1918 period would resist a man who "said it with flowers" – and although there is no message on the reverse of the card, the recipient probably knew from whom it came!

6 "Souvenir de Belgique": obviously dating from immediately after World War I, this rather lurid representation of the Allies' national colours was probably produced for sale to troops based in Belgium.

7 This Royal Engineers' card was purchased by a Sapper about to embark for France from an English port: his message on the reverse states that the card was on sale in several colour variations. Further, since it bears a message but is not stamped, we can deduce that these cards were normally sent through the post in protective envelopes.

8 The message on the back of this card, bearing the badge of the 15th (The King's) Hussars, reads simply: "From Uncle Jack, Aug 7th 1916". One wonders if the sender of this British-made card came home safely from the Front.

9 Made in Paris, this card of the Royal Sussex Regiment is unusual in that the upper badge section opens out to reveal a pocket in the lower part, containing a small card inscribed "To my dear Wife". It must have cost the soldier a fair portion of his pay.

10 "To My Dear Boy"; colourful and sentimental, this may have been sent by a Nurse to a soldier who had been under her care.

11 The badge of the Army Service Corps, forerunner of today's Royal Corps of Transport.

12 "Thoughts of Thee": another design of the type that suited the popular taste of the time.

13 "Souvenir from France": the inclusion of the flag of the USA on this card shows that it must have been produced in 1917 or later.

14 Royal Naval Air Service: made in Paris and, as may be seen, of superior workmanship.

15 Royal Welsh Fusiliers. This card is dated to the pre-1920 period by the spelling "Welsh": this was officially changed to "Welch" in 1920.

16 A purely decorative floral card with no message.

17 "To My Dear Sister". Family ties were closer in 1914-18 than now, as this card and those shown at (2) and (20) emphasize.

18 "With Best Love". Another Paris-made card with a small inset envelope – see (9) – for an additional message of a more private nature, perhaps.

19 Another variation on the badge of the Royal Army Medical Corps, see also (1). According to the reverse, the design was copyrighted in London but the card itself was made in France.

20 "To My Sister". Another "family" card – see (2) and (17) – incorporating suitable motifs: the family cottage (with a French "chateau-style" roof that may give a clue to the card's origin) and the forget-me-not flower.

"Sweetheart Collectables" of World Wars I and II

Like the examples shown on *pages 56-57, 58-59* and *186-187*, the items shown here fall into the category of "Sweetheart Collectables": that is, objects made specifically for purchase by servicemen as gifts for their loved ones. A good collection of such material can be put together relatively easily, since objects like these are still widely available, in good condition and at comparatively reasonable prices.

1 Silk handkerchief embroidered with a representation of a Royal Air Force officer's full-dress cap badge of pre-1922 type (ie, before the Garter emblem enclosing the motto was officially replaced by a simple circlet).

2-3 This silk handkerchief is embroidered with the name of Amman, which is now the capital of Jordan. Before World War II, when this memento of his service there was purchased by a British soldier, it was part of the British mandate of Palestine. The powder puff fits into a small pocket sewn into the handkerchief.

4 Embroidered postcard of World War I: a "family" message like some of the other cards shown here and on the preceding pages. Since these embroidered cards are fairly easily found, the collector may decide to specialise in cards on a particular theme: sentimental messages; regimental or unit insignia; Allied flags; or souvenirs of towns such as Ypres or Arras.

5 Embroidered silk handkerchief with a design that suggests it was purchased by a member of a Scottish regiment. See (9).

6 Scissors from a British Army issue sewing kit—*not* for embroidery, but for the soldier to "make do and mend" with his uniform and equipment! The "housewife" in which such items were kept is shown at (3) on *page 194*.

7 "Souvenir from France": aircraft in the British national colours above what is possibly a stylised representation of a Royal Air Force (RAF) unit crest. World War II items, like this one, are generally speaking somewhat more rarely found than World War I examples.

8 Powder compact bearing the cap badge of the US Army. The addition of the scroll with the word "Germany" to the badge suggests that this is probably a post-World War II item, purchased by a member of the Occupation forces.

9 The Scottish piper design obviously constituted a good selling line in embroidered silk handkerchiefs. It is to be hoped that this example, with the message "To My Dear Sweetheart", was not also purchased by the buyer of (5), "To My Dear Wife"!

10 Another handkerchief from Palestine, pre-World War II; this one embroidered "Bethlehem". See also (2-3).

11 Silk handkerchief embroidered with the badge of the Royal Artillery, dating from World War I.

12 British Army issue linen handkerchief, embroidered with a crown. This is the rank insignia of a major—but it may be thought that an officer of this rank could have afforded a souvenir handkerchief

of a less economical kind!

13 Silk handkerchief of World War I, embroidered with the flags of Britain and the Dominions – Australia, Canada, New Zealand – and France. The message, partly obscured in the photograph, is "A Present from France".

14-15 Further examples of the "family" and "sentimental message" style embroidered postcards of World War I: "To My Dear Mother" and "Forget Me Not".

16 Sometimes a badge will help the collector to date a specimen, as with this embroidered handkerchief bearing the badge of the Bedfordshire and Hertfordshire Regiment. Although the rebus (punning device) of the "hart" above the "ford" was a part of the badge from 1881, within the Garter insignia of the Bedfordshire Regiment of which the Hertford Militia formed a part, the full regimental title was not adopted until 1919.

Handbags issued to members of Allied women's services, World War II. (Above) Very much a utility item, this canvas bag with contrasting piping was issued to women of Britain's Auxiliary Territorial Force (ATS). (Below) A bag of good-quality leather, rayon lined, issued to members of the US Women's Army Corps (WAC). These were smart enough to be retained for use after the war by ex-WACs and are consequently now very rare, since most were carried beyond the point of no repair.

17 As with (16), the style of the badge of the Royal Armoured Corps enables this embroidered handkerchief to be dated to the pre-World War II period, before the adoption of the Corps' "mailed fist" insignia, emblematic of the hard knocks it gave the enemy.

18 Packet of needles from a British Army sewing kit; note the "broad arrow" marking that denotes a War Department issue item.

19 Thimble taken from a British Army sewing kit.

20 This spool of khaki thread was found in the pocket of a US Army combat jacket of World War II.

21 On this powder compact, dating from around World War II, the silk-embroidered Royal Air Force badge is given protection by a clear plastic cover.

British, French, German and US Respirators, 1939-1945

1 British Standard Service Respirator; issued to every British serviceman, 1939-45. There were several variations: most marks were like the respirator shown, with a separate filter canister on a corrugated rubber tube, but some had a smaller filter mounted direct on the outlet valve. The large disc on the "face" is a "voicemitter", which allowed the wearer to converse or use a communications set. The respirator was issued and carried in a canvas or cardboard container, with various anti-gas accessories, such as creams—see (8)—and special spectacles.

2 British Child's Respirator; intended for children between the ages of two and five years. It was almost universally known as the "Mickey Mouse", since the outlet valve in the lower-central "face" gave it a fancied resemblance to the famous cartoon character. Not many of these small masks now survive in good condition.

3 Anti-Gas Ointment, No 2; issued to British servicemen and civilians, World War II. The ointment was rubbed into the skin to give a protective coating against blister chemicals, or to clean off any traces of them. The evil smell of some of these ointments made them much disliked by those who had to use them in anti-gas practice.

4 French-type Model 1939 Respirator; something of a mystery—a respirator of French pattern but with internal markings in Serbo-Croat. The plastic moulding that holds the metal filter in place incorporates a plastic "voicemitter".

5 US Service Respirator, 1941; the type dates from 1918. Probably the ugliest and clumsiest of all service respirators, this US model was heavy and uncomfortable. The face-piece is of linen-covered rubber and the long corrugated tube terminates in a large filter canister that was worn in a satchel on the chest. The outlet valve is a

Baby's Protective Helmet: the official designation of the gas-mask issued free of charge by the British Home Office in World War II to all families with children under two years of age. Infants could not inhale through the filters on conventional masks, so filtered air was pumped into this helmet by the bellows at the side.

simple rubber flap, protruding downwards from the same moulding as that used by the tube. Issued in tens of thousands, these US respirators may still be found quite easily.

6 British Special Services (SS) Respirator; World War II. Issued to such civilian services as firemen, ambulance crews and civil defence units, this model was light and handy, since the face-mounted filter removed the need for a long, clumsy tube. However, there was no provision for voice transmission other than a microphone connection. Normally carried in a shoulder-slung canvas satchel.

7 British Anti-Gas Handbook; several versions were issued in the years up to 1939, for the information of the various civil authorities who would be responsible for defence in the case of gas attack. As shown, these included Britain's police forces.

8 Anti-Dim Ointment; usually carried in the respirator case by British servicemen and civilians, this cream was intended to reduce misting of the eye-pieces by water vapour from the wearer's breath and body heat.

9 German GM 30 Service Respirator; World War II. The Germany Army was well-prepared for chemical warfare and the respirator available in 1939 was both well-designed and of excellent materials. The combined filter-canister/voice-mitter is high-quality dural moulded and the face-piece is of rubberised fabric. These now fetch high prices in good condition—especially when complete with canister (10)—and the example shown is given added value by having the owner's name, "Müller", and service number on face-piece and strap. The type is still in service in East Germany and Czechoslovakia.

10 Canister for (9), of fluted grey-green metal. This was always worn with German Army combat dress, as contemporary photographs show; but since it was often used to carry rations or personal possessions rather than for its proper purpose, it is rare to find one complete with respirator.

11 British General Post Office (GPO) Telephonist's Respirator; World War II. Essentially an SS model—see (6)—but with the addition of a headpiece for earphones and a microphone connected to a GPO switchboard plug, this was issued to GPO personnel who might have to man essential switchboards during a gas attack. A rare item.

12 Special Adhesive: an unusual item dating from World War II. This special compound was intended to seal cracks in windows and doors against gas; it could also be applied to windows to protect them from blast damage.

13 British Civilian Gas-Mask; World War II. These were produced in millions and were issued to all civilians, but so many were discarded immediately after the war that they are now becoming rare and hard to find. They were usually slung from the shoulder in stout cardboard containers: a gas-mask with its box would now be an excellent find for the collector.

Miscellaneous Souvenirs: World War I to the Falklands

1-2 This large-capacity water bottle, blanket-covered and with leather loops for a carrying sling, bears the British Air Ministry stamp and the King's Crown mark and was probably used in pre-World-War-II RAF operations in the Middle East. Its lid, or cap (2), forms a drinking mug.

3 This china mug with its motif of bombers and submarines over and in the Atlantic Ocean commemorates the Anglo-American Alliance of World War II; the handles are formed by figures of US President Franklin D. Roosevelt (left) and British Prime Minister Winston Churchill (right).

4 The immense cost of fighting World War II led to official implementation of various schemes to encourage citizens to save—and to lend their savings for the war effort. The major British scheme consisted of National Savings Certificates: these could be bought by instalments in the form of stamps, which were stuck into the book shown here.

5 A set of bottle stoppers in the form of the busts of Anglo-American leaders at the time of victory in 1945. Left to right: US President Harry S. Truman; General Douglas MacArthur, USA; General Dwight D. Eisenhower, USA; Prime Minister Winston Churchill, Great Britain.

6 Mail between friends in World War I: a Christmas card sent by an officer in the 28th London Regiment (Artists Rifles) to an officer in a Trench Mortar Battery.

7 Cigarette case made from scrap aluminium by an Italian prisoner-of-war for a British NCO in the Middle East; note "eastern" motifs.

8 Souvenir of modern war: a Zippo cigarette lighter commemorating the Falklands Campaign of 1982.

9 This small hand bell, sold to raise money for the Royal Air Force Association during World War II, is made from the metal of German aircraft brought down over the UK. It bears the "Victory V" on the handle and a profile portrait of the Soviet leader Josef Stalin.

10 Bags like this, containing sweets and gifts to celebrate the Feast of the Passover, were distributed to members of the Jewish Brigade serving with the British Army in World War II.

11 A fly switch in the traditional African style sometimes formed a necessary part of the equipment of European officers in North Africa.

12-13 Officially produced during World War II, these wooden models in 1/72 scale were used to train personnel in aircraft recognition. Those shown are British naval aircraft: (12) Fairey Firefly; (13) Fairey Barracuda.

14 British housewives who gave up their aluminium pots to be used in the construction of Spitfires in 1939 sometimes received souvenir compensation: as with this very fine china teapot.

16 Badges issued during World War I to defence and munition workers who otherwise might be branded as "shirkers", because they were not in uniform and perhaps handed the "white feather" emblem of cowardice by some over-patriotic lady.

17 Royal Air Force issue drinking mug

A souvenir of Imperial Germany: a child's uniform tunic (shown with the "attila" jacket of a lieutenant of a Brandenburg cavalry unit, c1900) with parade epaulettes of Baden Life-Dragoon Regiment No 20. Officers had such tunics made for their sons to wear on festive occasions.

of pre-1939 period, used in ORs canteens on permanent stations.

18 Victory commemorative saucer of 1920: for units that served in the anti-Bolshevik intervention in Russia, the war went into 1919.

19-22 These lapel badges were worn by official house-to-house collectors for National Savings in World War II in Britain; their years of service are shown on the bar.

23 Some soldiers could take into civilian life skills acquired during their service: this British War Department Driving Permit could be exchanged for a civilian licence.

24 "Black-Out!" – a contemporary card game based on the British air-raid precautions of World War II.

25 Stoneware tankard bearing the emblem of the National Fire Service – heroes of the "Blitz" of 1940-41.

26 Some Roman Catholic civilians in Britain carried rubber tags like these in case of injury in air raids.

27 This white metal model of a German V-1 ("Doodlebug") flying-bomb was made by a US serviceman as a present for a British family in 1945.

28-29 Tunic button of the Metropolitan Police, London's "regular" police force, and (29) a temporary armband of the Special Constabulary, for which extra members were recruited in wartime.

30 Many packaged items – especially cigarettes and matches – were produced in "information" packs in World War II: these Sweet Caporal cigarettes, with aircraft recognition data on the pack, were issued to Canadian forces.

31 Books were specially printed for servicemen overseas: this "western" pocket-book was issued free to US servicemen.

32 Commercially-produced patriotic matchbox covers of World War II (above) and World War I (below).

33 Silk squares showing, in this case, cap badges of the British Army, were given away with cigarettes in the World War I period; perhaps a heavy smoker produced this very attractive and colourful contemporary cushion-cover.

1 2 3 4 5 6 7 8 9 10

1 The winter of 1939-40, the first of World War II, was exceptionally severe, and a call went out on the "Home Front" for "Comforts for the Forces". Patterns were produced for items suitable for all the fighting services—in Air Force blue or Navy blue for the RAF and Royal Navy; in khaki for the Army—and women at home or in groups organised by such voluntary organisations as the Women's Institute produced vast quantities of knitwear for the forces.

2 An example of the home knitters' skill: mittens produced from a pattern requiring some 2oz (57g) of four-ply wool.

3 A "housewife"—a folder of olive drab material in which a serviceman kept a needle and thread, spare buttons, and other items for repairing his kit—issued to a man of the US Army Air Corps.

4 US Government Issue (the abbreviation of the phrase is sometimes said to be the origin of the term "G.I." for an American soldier) shaving set: a Gillette safety razor with spare blades.

5 For use with (4), perhaps: a soap shaving stick by Erasmic of London, in wartime packaging.

6 An officer's riding crop with a silver top, bearing the badge of the Royal Artillery, and a leather-bound handle.

7 Recreation for the troops in World War I: a pack of playing cards emblazoned with the flags of the Allies (including Imperial Russia) and showing the British bulldog and French poodle savaging the German flag. Packs like these were distributed free to wounded and sick servicemen in hospital.

8 The women at home were busy with their knitting needles in World War I, just as in World War II (see 1 and 2). The Balaclava helmet—this example dating from 1914-18—was a popular pattern in both World Wars.

9-10 Brass "button sticks" of the World War I period. These were slipped over buttons, buckles and badges while they were being cleaned, in order to prevent metal polish from soiling the uniform. To many present-day servicemen, with "stay-bright" buttons, these must truly seem museum-pieces!

11 A presentation cigarette box, silver-plated, bearing a relief based on the work of the famous war artist R. Caton Woodville and showing British troops overrunning a German field gun battery early in World War I. It bears also a quotation from Rudyard Kipling, whose patriotic verse was then at the height of its popularity.

12 Any former soldier will recognise these materials with a sinking heart! Renovatory polish for webbing equipment: (below) "Blanco" of the World War II period; (above) "Pickering", which replaced "Blanco" in the British Army during the 1950s.

13 The British soldier of World War I kept his personal kit—like the cut-throat razor, shaving brush and fibre button stick shown; all marked with the owner's service number—in this linen roll.

14 Commercial souvenir in pewter: an RAF pilot of the Battle of Britain

period, 1940, runs for his aircraft.

15 Cool wear in the Pacific, World War II: a Japanese soldier's loin-cloth.

16 Japanese-style "comforts for the troops": fans were sent from home to men serving in the tropics.

17 Silk head scarf of World War I: the flags of Britain, the Dominions and the Allies frame a patriotic poem by Harold Begbie, first published in the London *Daily Chronicle* in 1915.

18 These unusual and attractive playing cards (compare with 7) were produced by the Coca-Cola Company for use by US servicemen during World War II. Aircraft recognition silhouettes are shown on each card, along with conventional values.

19 A tragic "souvenir": a plaque sent officially to the next-of-kin of British servicemen who fell in World War I.

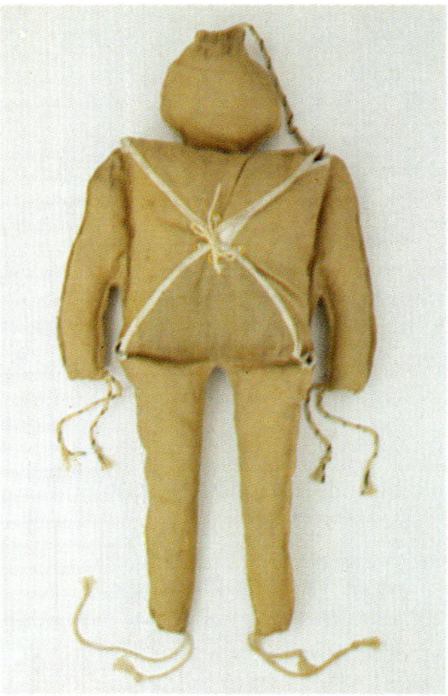

Dummy, about half-human size, of hessian filled with sand, with an integral parachute. These were air-dropped inland from the Normandy beaches on the night of 5-6 June 1944, to facilitate the Allies' "D-Day" landings by creating alarms of parachutists.

20 An ash-tray made from the piston of a Rolls-Royce Merlin aero engine—used in the Spitfire and Hurricane fighters—and inscribed with Winston Churchill's famous words on the Battle of Britain pilots. These were sold by tobacconists during World War II to raise money for the RAF Association.

21 Empire Day (now Commonwealth Day), 24 May, was used during World War II as a major fund-raising occasion for comforts for servicemen from the British Commonwealth. This certificate was awarded to one Alan Robinson to mark his efforts in 1941.

22 "Victory V" book matches, produced for sale in service canteens, often had useful "tips" printed on the reverse of the folder.

23 A topical gift of 1939-45: a glass bottle to hold bath salts or oil, in the form of an RAF pilot.

24 These playing cards were produced for the recreation centres run by the Knights of Columbus fraternal society for Canadian servicemen in World War II.

25 Tin-plate picture frame with an acetate shield: the card within showing scenes of British servicemen in action provides a suitable background for a soldier of World War I.

Foodstuffs and Eating Utensils of World Wars I and II

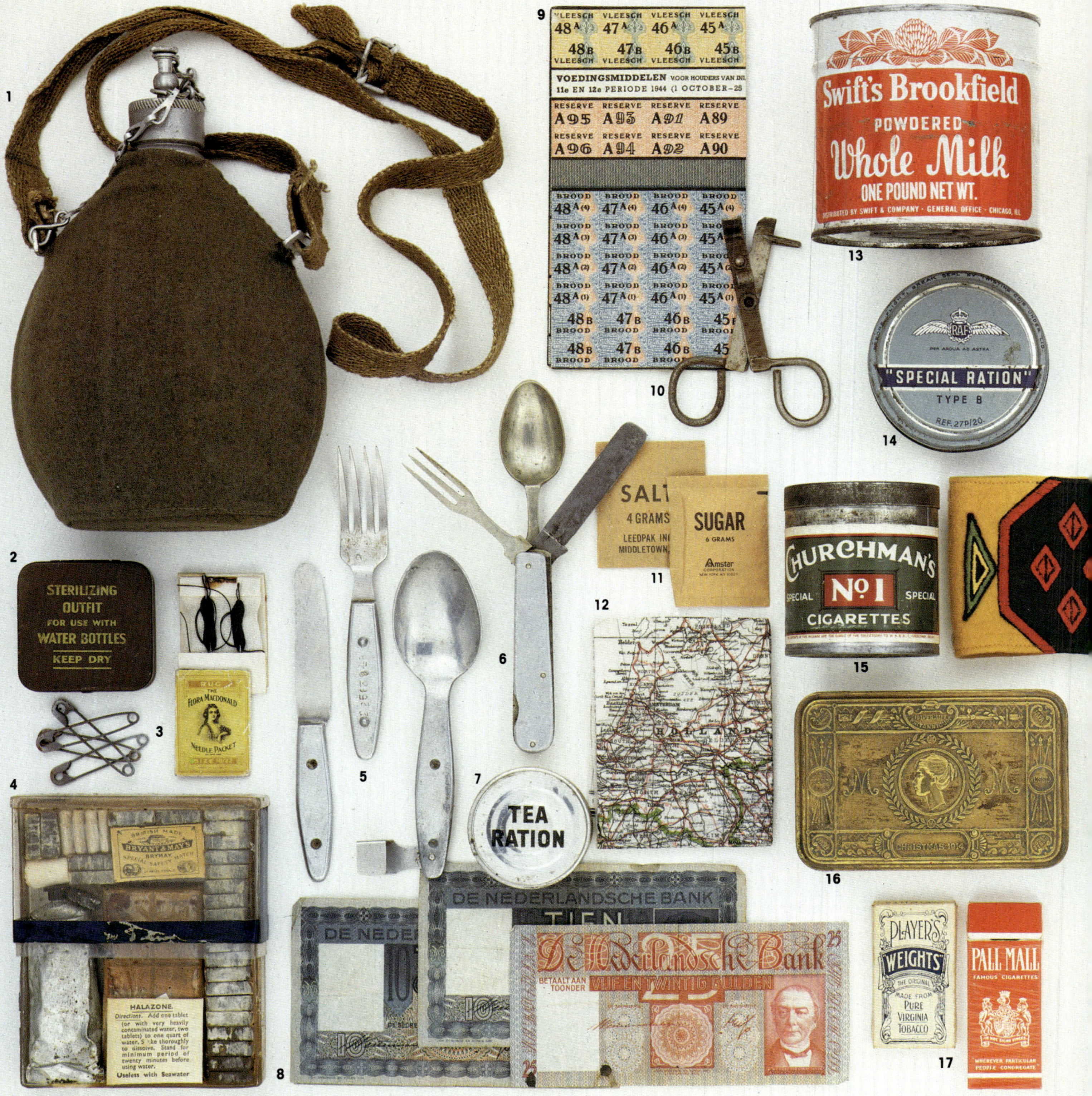

1 This Italian Army water-bottle of the World War II period is extremely strongly constructed of fluted aluminium, covered in blanket material. Its screw stopper is secured by a chain. The shoulder-sling, however, is of poor quality.

2 A tin of tablets issued to British Army personnel for sterilizing metal water-bottles before refilling, World War II.

3-4 The fishing hooks and line, safety pins and needles shown at (3) are from the Royal Air Force issue emergency pack (4), for the use of aircrew brought down in enemy territory or in the sea. The pack contained more than a dozen potential life-saving or life-sustaining items in a waterproof plastic container, including matches, food tablets, anti-sunburn cream and water-purifying tablets; as well as currency (see 8) and a map (see 12) relating to the area over which the aircraft was operating. The contents varied, but they invariably included a magnifying glass (for making fire), a rubber water-bottle, food tablets and chewing gum. (See also 30.)

5 British Army issue knife, fork and spoon of World War II. Made of stainless steel, with alloy handles, all three clip together for carrying. Owner's service number is on fork.

6 The same utensils as (5), but dating from World War I and hinged together.

7 British Army issue tea ration of World War II: there is probably enough tea in the sealed tin to make about 1 gallon (4·5 litres).

8 Dutch currency taken from the RAF emergency pack (see 4).

British Army thermos flask, used to carry hot food to forward positions. Made by the Thermos Company in 1943, it has an insulated and padded carrying container.

9 These are forged Dutch food coupons which were issued to Allied airmen who might be forced down in enemy-held territory.

10 An invaluable aid to the British grocer during World War II: the clippers are specially designed to cut coupons from ration books and retain them on the spike seen on the right-hand side.

11 Salt and sugar packets from a US Government Issue ration pack.

12 Silk map of Holland issued to RAF bomber crews (see 4).

13 Dried milk supplied by the USA to Britain under the Lend-Lease Act of March 1941. Basically, this meant that Britain leased bases to the USA (then a non-combatant) in return for material aid.

14 One of the enigmas that will face all

militaria collectors from time to time: is this tin a special "Type B" ration issued to British aircrew – or a wartime brand-name for a commercial tobacco?

15 Churchman's No 1 cigarettes packed in a special tin of 50 for sale by the NAAFI (Navy, Army and Air Force Institute) in its canteens for British servicemen, World War II.

16 Each man of the British Expeditionary Force (BEF) in France at Christmas 1914 was presented with one of these finely-made brass boxes, containing chocolate, cigarettes and a Christmas card – a gift from Princess Mary.

17 Small cigarette packs of the World War II period: Weights, a British brand; extra-long Pall Mall, specially packed for inclusion in US forces' combat rations.

18 Standard US Government Issue mess tin, knife, fork and spoon. The tin, made of heavy-gauge stainless steel had a lid (not shown) held in place by the folding steel handle when stowed in the pack.

19 Armband of the Women's Land Army, formed during World War II to train women in agriculture.

20 Breast pocket badge from the white overall of an employee of the NAAFI (see 15).

21 NAAFI badge from a woman kitchen worker's overall; the NAAFI employed a maximum of 60,000 women in 1939-45.

22 Bronze badge worn by a voluntary worker in the British Army Canteen service, World War I.

23 Band and badge from the headgear worn by a waitress in an Officers' Mess club run by the NAAFI during World War II.

24 Simple tin-opener supplied with all US field packs containing tinned food, 1941-45.

25 China tea in packaging typical of pre-1939 and early wartime Britain.

26 Another typical Lend-Lease (see 13) commodity: US-made dried eggs.

27-28 Individual ration tins from US Government Issue field packs.

29 Many NAAFI items were specially packaged in marked tins – like this cigarette tin – to prevent unauthorised sale to civilians.

30 RAF survival pack (as 4), shown here unopened.

31 Sealed Survival Ration tin, issued for equipping the inflatable life rafts of USAAF and USN aircraft.

32 Canned butter concentrate for use in the tropics, World War II.

33 Emergency Flying Ration issued to RAF aircrew in Burma and Malaya.

34 Instant coffee packs from US Government Issue combat rations.

35 Many an infantryman must have carried these: corn caps for feet suffering the effects of long route marches!

36 A complete meal sealed in a tin: issued to Allied forces in forward areas, Pacific Theatre, 1941-45.

37 Toothpaste in World War II packs: (above) Gibbs Dentifrice, a popular brand for British civilians; (below) the same item packaged for sale aboard H.M. Ships.

38 Ovaltine, a British chocolate drink, in pre-World War II packaging. Extensively used in forces' convalescent homes, it was in short supply on the "Home Front".

39 More dried milk for wartime Britain, from the Dominion of Canada.

Red Cross and Medical Items of World Wars I and II

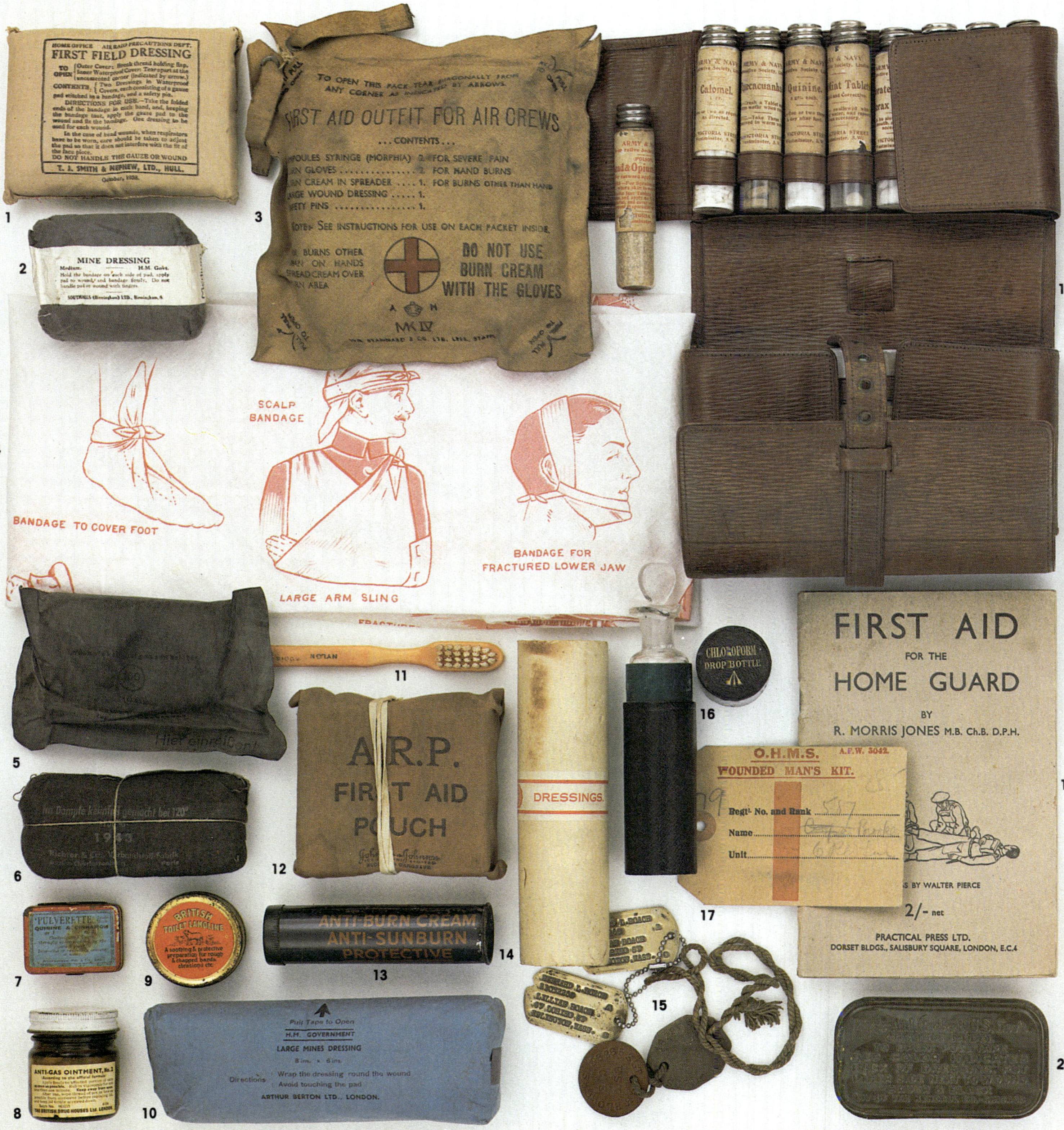

1-2 British wound dressings, World War II: (1) ARP (Air Raid Precautions) dressing for wounds caused by flying glass in air raids; (2) a service issue dressing for more serious wounds.
3 Waterproof pouch carried in RAF aircraft and containing material to deal with serious wounds.
4 Sling, printed with instructions for use, produced in World War I.
5-6 Germany Army field dressings of World War II, in waterproof wrappers. Note that (6) was made in Paris.
7 Proprietary brand of quinine tablets for protection against malaria.
8 British Anti-Gas Ointment, applied to the skin to counteract the burning effect of mustard gas.
9 This ointment was included in Civil Defence first-aid kits in the 1940s, for use on the hands of air raid heavy-rescue workers.
10 British service issue dressing, World War II. (See also 1-2.)
11 British service issue toothbrush, made in 1944 of a then novel material—nylon.
12 British ARP worker's First Aid kit, sealed in a waterproof pouch.
13 Anti-burn cream: included in the stores of RAF inflatable survival rafts, against exposure.
14 Lint from an ARP First Aid box.
15 Allied troops' identity discs of two wars: (above) stainless steel "dog tags" of US forces, World War II; (below) fibre discs of the British Army, World War I.
16 Chloroform drip bottle, for anaesthesia in advanced dressing stations, World War I.
17 Label attached to wounded man's kit sent from forward area, 1914-18.
18 Leather-cased drug kit for tropical use by British Medical Officer.
19 First Aid Handbook for British Home Guard units, 1940.
20 US soldier's personal First Aid Kit, hygienically sealed in a tin.
21 First Aid Kit carried on some US vehicles, World War II: a comprehensive kit in a waterproof metal container.
22 Sterile burn dressing issued to British Civil Defence posts, c1940.
23 British Army issue medicated talcum powder for feet and body.
24 Proprietary aspirin from a British household's first aid kit, 1939-45.
25 Post-World War I shoulder title of the Red Cross and St John's Ambulance Association: the two organisations were amalgamated in both World Wars.
26-29 Red Cross items: (26) drinking cup for reclining patients; (27) badge from nurse's apron; (28) armband of World War I, giving the wearer protection under the Geneva Convention; (29) soap supplied by Canadian Red Cross, probably to PoWs.
30-32 US issue medical supplies: (30) sulphur drug, widely used before the advent of penicillin; (31) boxed dressing and tourniquet; (32) items from US Army First Aid Emergency box.
33 British Civil Defence dressing.
34 Armband issued by Germany to give medical orderlies the protection of the Geneva Convention. This example was worn by a British PoW, 1939-45.
35 US Army first aid pouch, 1942, worn on the ammunition belt.

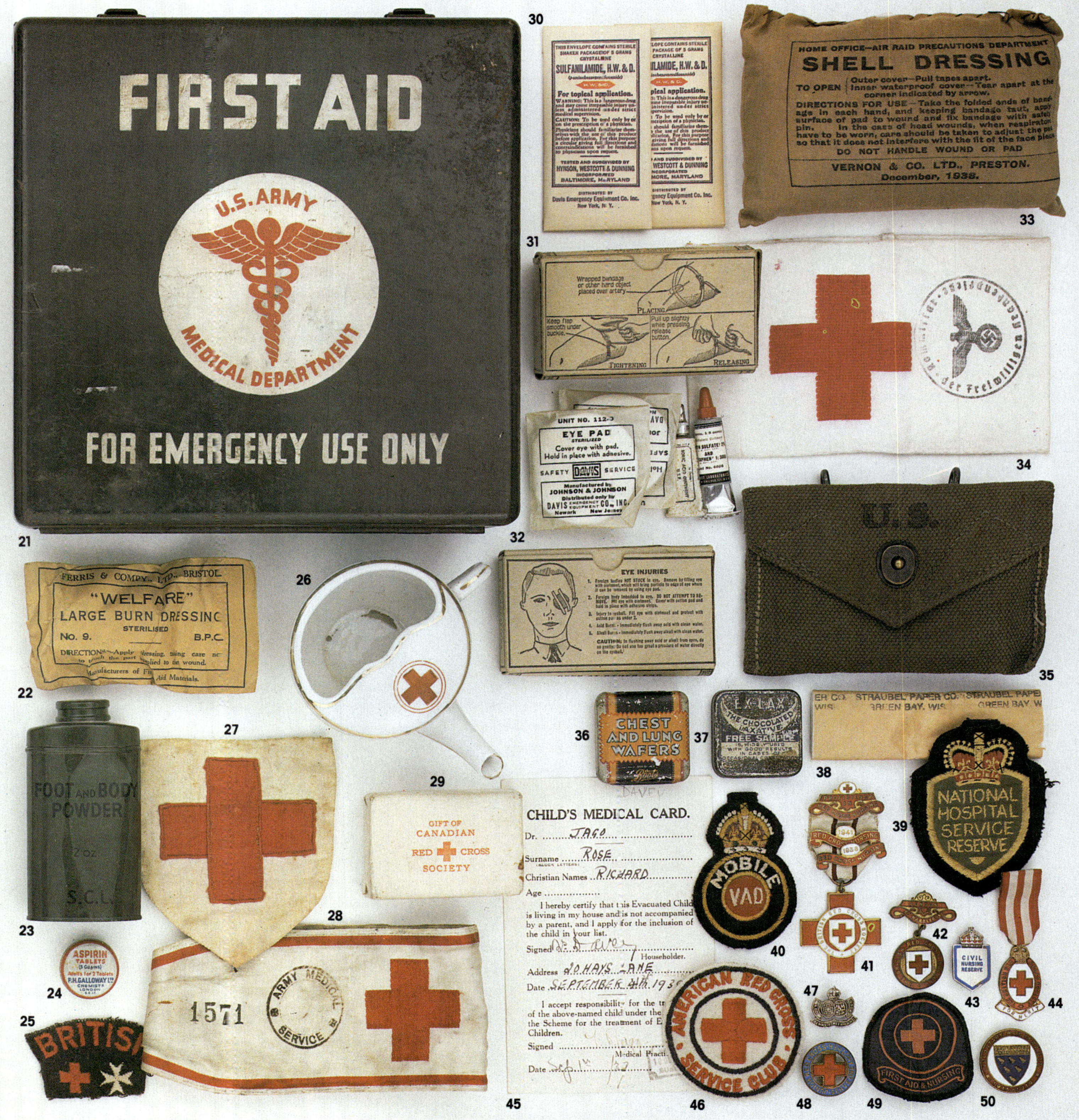

36-37 Proprietary remedies of 1939-45: (36) treatment for bronchial ills; (37) a popular brand of laxative.
38 Toilet paper from a US combat ration pack, World War II.
39 British Hospital Service badge, worn by non-medical staff.
40 Voluntary Aid Detachment (VAD) badge, for drivers of ambulances, mobile canteens etc, World War II.
41 British Red Cross medal for proficiency in nursing, 1938-1941.
42 Pendant badge worn by a Red Cross nurse at a London hospital, 1938-45.
43 Lapel badge of the Civil Nursing Service, World War II.
44 British Red Cross medal for meritorious service, 1936-52.
45 Medical card of a child evacuated from an air-raid threatened area of Britain, only two days before World War II began.
46 Badge of the US Red Cross branch that administered social clubs and canteens in Europe, 1942-45.
47 Lapel badge of savings' organisation of British Red Cross/St John's Ambulance amalgamation.
48 Membership badge of the British Junior Red Cross Society, open to children aged 6-14 years.
49 Overcoat shoulder title of British Red Cross worker qualified in First Aid and Nursing.
50 Lapel badge awarded after World War I for voluntary service at a British hospital in 1915-19.

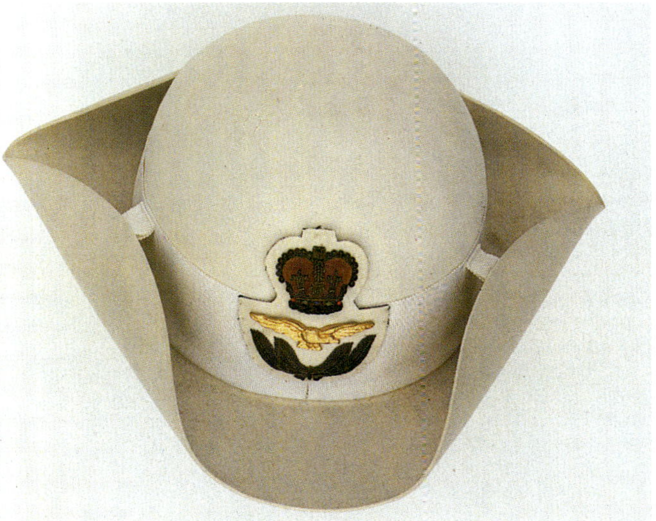

Hat issued for Middle East service to officers of Princess Mary's Royal Air Force Nursing Service; formed in 1918 and a permanent branch of the RAF since 1921.

Binoculars, Maps and Compasses of World Wars I and II

1 German officer's field glasses, World War I. Many officers had to provide their own field glasses; thus commercial models were in service and examples with military markings are rare.

2 British Army issue binoculars, World War II. Most officially issued binoculars, of which these are typical, were 7x32 or 7x40. They may still be found quite easily.

3 US Army Air Force issue sun glasses. These were prized acquisitions during World War II and many are still in use. A pair with the original case would be a very worthwhile find for the collector.

4 British gas-mask spectacles, World War II. The close fit of the official issue gas masks (see *pages 190-191*) meant that spectacle wearers had to obtain specially-designed glasses to wear with them. Expensive then; rare—and expensive—now.

5 Map of North Africa, World War II. In the early part of the war military maps were in short supply and had to be supplemented by commercial issues like this one. Maps usually get hard usage, so wartime examples in good condition are highly collectable.

6 The import of the title of this US Forces map is grim: the second atomic bomb attack was made on Nagasaki, 9 August 1945. Any war map with such connotations is eagerly sought.

7 US Army Air Force Aerial Dead Reckoning Computer. Navigation instruments are always in great demand, but copies of such

Map case and instruments of an officer of the Royal Artillery, World War II. This rare item was found complete with code books, pencils and annotated map, seen here.

items as the one shown here—which is still in use—should be guarded against.

8 Royal Air Force crews of World War II carried photographs of themselves in civilian clothes when on missions over German-occupied Europe. If they were forced down, these could be used for the production of fake identity papers like the example seen here. Such items are now rare.

9-11 More escapers' aids: aircrews were often issued with compasses concealed in their uniforms, usually in buttons, as shown. Note also RAF locket compass.

12 Combined map case and navigation board of the kind made for British cavalry officers in World War I and used also by the Royal Flying Corps.

13 Clock/Stopwatch from a Messerschmitt Bf 109. When an aircraft was brought down, such small and attractive items were always the first to "vanish".

14 Royal Artillery slide rule, 1918. These may still be found complete with cases and accessories.

15 British trench periscope, World War I. These were produced both commercially and by the services: both varieties are now rare and fetch high prices.

16 Aircraft recognition cards issued commercially by *Flight* magazine, World War II.

17 Waterproof map case issued to British aircrew, World War II.

18 Presentation plaque showing the British Occupation Zone of Germany; examples like this from the immediate post-World-War-II period are quite rare.

19 Map of the Middle East, dating from World War I.

20 Silk escape map covering part of the Balkans. These were issued to aircrew in World War II and are not rare.

21-22 British issue marching compasses. A collection of compasses of as many different types and nationalities as possible would be attractive—if somewhat expensive to assemble.

23 "Tinned" heliograph mirror, World War I; found in several different designs.

24 Duralumin War Knife, 1914; one of many commercial products rushed out for "war service" at the outbreak of World War I. They rarely lasted long in use and are now rare.

25 Map of the Naples area, 1944; cut from a larger map and linen-backed, probably for use by a service transport driver.

26 British marching compass, World War I. The value is much enhanced because it is still complete with leather case.

27 Commercially-produced lady's headsquare, 1939-45. Beware of such items purporting to be escape maps.

28 RAF issue Dinghy Knife and case. These light knives with dural or thin steel blades and cork handles were produced in large numbers.

29 Miniature compass; part of an emergency navigation kit issued to RAF aircrew for use after crashing or ditching. *Not* an "escape" item and not rare.

British, US and German Signals Equipment, 1914-1945

1 This small map case, dated 1944, was obviously intended for use in the Far East: it is made in jungle green and bears a warning concerning mosquito repellent. British tropical kit is fairly rare: the life of webbing equipment was short in jungle conditions and few items were brought back as souvenirs at the war's end.
2 Storage tin and, to the right, mirror, of the British Army's long-serving Mk V 5in Heliograph. The mirror was set up on a tripod and, depending on the strength of the sun, was used to signal in Morse Code flashes over a number of miles. Light signalling of course, has persisted into the electronic age, principally in ship-to-ship communications.
3 Standard signalling pistol of the Luftwaffe, of alloy with plastic grips for lightness, with Nazi eagle stamps (below the chamber) and dated 1940. Firing coloured stars, it was used for visual signalling of all kinds. A highly-prized collectors' item.
4 This waterproof red light, issued to RAF aircrews in World War II, was intended to be clipped to the "Mae West" lifejacket or to the inflatable dinghy used if the aircraft was forced down in the sea. There was a good chance that the continuous or flashing red light would help an air-sea rescue aircraft or launch to locate men adrift in the water.
5 Standard angled torch, olive drab in colour, usually over brass, issued to US Forces in World War II. It was provided with coloured lenses for use in traffic control.
6 An interesting item from the German occupation of the Netherlands, World War II: a metal self-generating torch made by Philips. A pumping action of the toggle on top drives a small generator to produce enough current for a 2·5 volt bulb. Dry-cell batteries were almost unobtainable under the occupation and, since blackout regulations were severe, torches like this were valued.
7 Based on the bicycle lamp of the period, this was the most common type of torch issued to the British Army in World War II. Of sturdy metal construction, it has a movable visor over an opaque lens glass, with a movable red half-lens for use in traffic control. The visor shields the light from

Cloth badges of the Royal Navy Electrical Branch, 1955-79 period. The letters indicate Radio Electrician (R) and Ordnance Electrician (OE).

above, in accordance with blackout regulations.

8 Much rarer than (7), this British Army lantern of World War II consists of a metal frame holding a 4·5-6 volt dry cell battery and a nickel-plated front container incorporating switch, bulb holder, reflector and glass.

9 Cloth armband worn by men of the Royal Signals, British Army, for identification in the field; World War II.

10 A very handsome piece of signal equipment dating from 1915, this morse key of brass and mahogany has a similar box (not shown) for storage. Items like this are fairly rare, probably because a great deal of brass was salvaged as scrap.

11 The famous US Army "walkie-talkie" of World War II, the BC-611-c radio that allowed small units to stay in touch with HQ. This sophisticated piece of equipment had great influence on the development of both military and civilian radio communications.

12 German Army electric signalling lamp of World War I, of heavy gauge metal with nickel-plated lens holders.

13 British Army "Lantern, Electric, Traffic, No 2" of World War II: weatherproof, with a push-down brass on-off switch in the top cover; with small "feet", so that it could be placed on the ground clear of moisture; and powered by the battery seen at (19).

14 British Army "Torch, Signalling, Mark III" of World War I – of inferior manufacture, but with an "envelope" on the front for storage of the red acetate lens when not in use. The lug (left) accepts the 1907 Pattern bayonet, enabling the lamp to be positioned clear of the ground.

15 This British-made electric lantern of the 1930s was not intended for military use, but would be a worthwhile addition to any collection of torches and signalling lamps.

16-17 Made by the wellknown British military supplier Gale & Polden of Aldershot, these symbols are mounted on magnets and are intended for use on a sheet-steel blackboard in planning military operations.

18 US Forces "Pyrotechnic Pistol AN-M8" of World War II. Lugs on the barrel allow it to be locked into a port for safe firing from a confined space – such as the interior of a B-17G Flying Fortress, in which this example was carried.

19 Standard double dry cell battery of British Army lamps and lanterns, World War II. Not easy to find, since if they are not kept dry they have a tendency to "sweat" and dissolve.

20 Small, flat, dry cell battery dating from the British "blackout" of World War II, when a torch was a necessity and batteries were in short supply.

21-22 More batteries of the same period, the larger one of Canadian manufacture.

23 Standard message pad of the Royal Signals, British Army, with a refillable cover of tough, waterproof plastic.

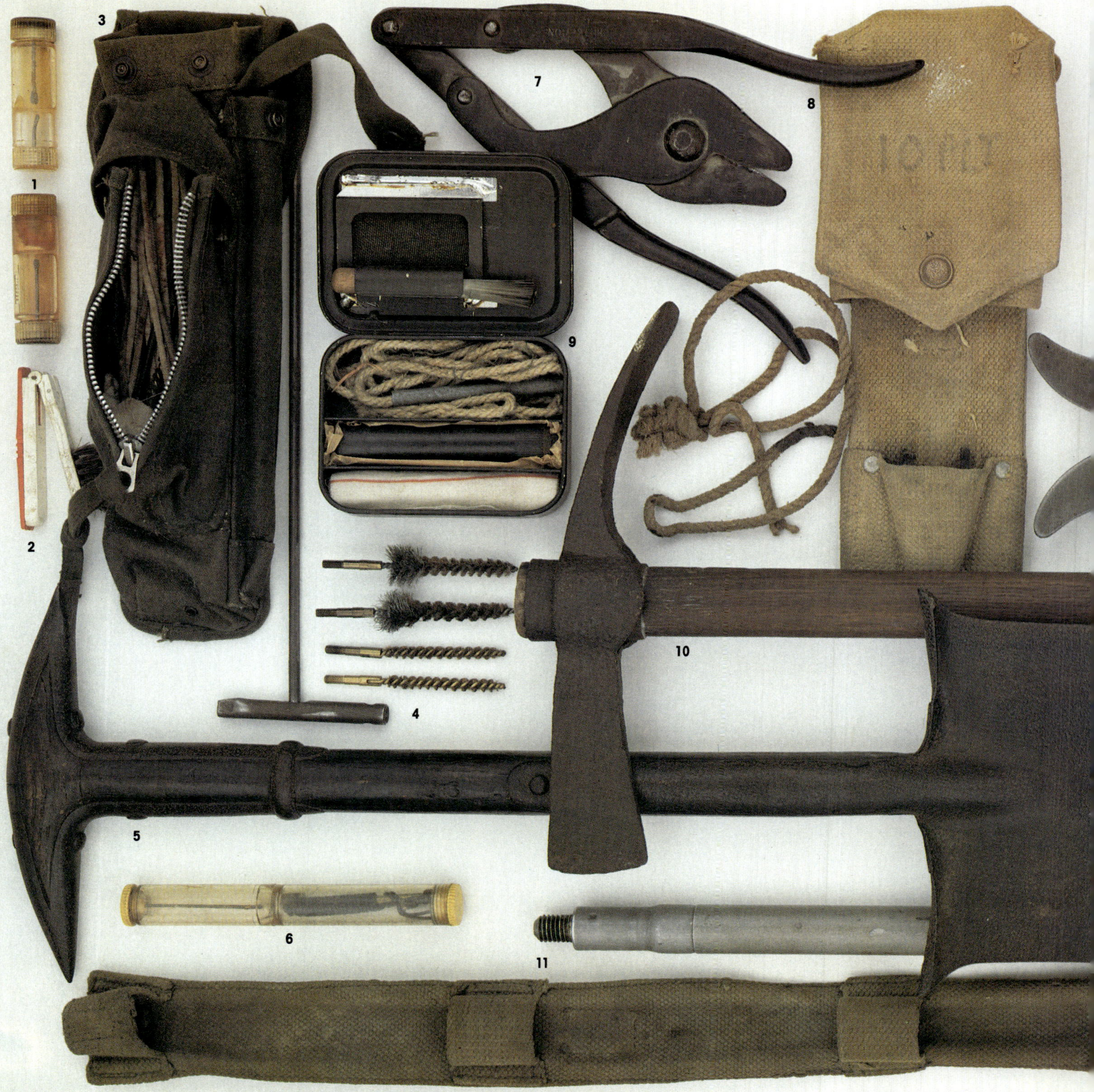

1-4 US Army issue complete cleaning and field servicing kit for the ·30 calibre Browning Model 1919A4 and ·50 calibre Browning M2 machine guns, both of which were extensively used by infantry units and on armoured fighting vehicles in World War II and later. The M2 remains in service with US forces and with some 20 other countries; the M1919A4 is obsolescent. The outfit shown here comprises (1) oil bottles; (2) folding brush; (3) carrying case in olive-drab canvas, with zip-fastening; (4) a steel cleaning rod, the handle of which forms a disassembling tool with probe and socket wrench, and a set of cleaning brushes that screw into a threaded socket on the end of the cleaning rod.

5 Ever since the American Civil War of 1861-65 demonstrated the awesome fire-power of the breech-loading rifle and field artillery with improved ordnance, the entrenching tool has been one of the infantryman's basic survival weapons. The example shown is a British Army tool dating from the Boer War, 1899-1902: to a tough wooden shaft are fitted at one end a spade and at the other a pick.

6 Plastic combination container for an oil bottle and pull-through for the US ·30 calibre Rifle M1 Garand, in service with the US Army from 1939 to 1957 (it is estimated that total production of Garands amounted to around 5·5 million). This small cleaning kit was carried in a recess in the butt; it is plastic to avoid adding significantly to the unloaded weight of 9·48lb (4·3kg).

7 Appearing on the battlefield in the later 19th Century, wire entanglements necessitated another addition to the soldier's equipment. These British Army folding wire-cutters date from World War II.

8 Webbing carrying case for (7), with the stamp of "10 PLT" (10 Platoon) of an infantry company.

9 Improvised British cleaning kit for the US-made ·45in Thompson Sub-Machine gun: the tin contains cleaning brush, pull-through, oil bottle and "four-by-two" cloth; the last-named being so called because of the dimensions of the pieces that were torn off to be "pulled-through" the barrel. Although the first model of the "Tommy Gun", named after its inventor, Brigadier-General J.T. Thompson, USA, appeared in 1921, it was not adopted for US Army service until 1938. Britain, which had not developed a standard sub-machine gun (some traditionally-minded staff officers, it is said, opposed the introduction of these "gangster weapons"), hurriedly purchased a large number of Thompsons in 1939-40. Total production, ending during World War II, is estimated at around 1,400,000, of which number a few were made in Britain by Birmingham Small Arms (BSA). The cheaply-produced and somewhat crude British Sten sub-machine gun which gradually replaced the Thompson in British service was nicknamed "the tin tommy gun".

10 British Army entrenching tool of World War I: one of several types

issued for use in the construction of the vast system of trenches on the Western Front. As may be seen, it also made a handy weapon for close-quarter combat in such "silent" operations as night raids on enemy trenches!

11 British extensible mine probe of alloy, with webbing carrying case. This tool was produced for use by the Royal Engineers towards the end of World War II—and found use again in the aftermath of the Falklands Campaign of 1982, when it was used to locate Argentinian plastic mines.

12 US Army brass tyre pressure gauge, calibrated to 160psi and designed for testing heavy twin-wheeled vehicles. It bears the stamp: "US Property Ordnance Department".

13 US Army issue camp axe, World War II. This was normally issued with a canvas carrying case, with hooks for attachment to an ammunition or pistol belt.

14 British axe of high-quality steel with a heavy rubber insulated handle, carried in aircraft of the Royal Air Force, World War II, to enable crews to cut their way out of a crashed—and perhaps burning—aircraft. Also issued to the Fire Services and Civil Defence units, since it could be used to cut through power cables in an emergency situation.

15 Wire cutters of World War II; manufactured by a British firm, 1918, but very similar to the French Army pattern of the period.

16 Universal Tool issued to US servicemen equipped with the famous ·30in M1903 Springfield Rifle—see *pages 150-151*, (6)—which was in service with US forces in both World Wars.

17-19 Handles (17), blades (18) and maintenance tool (19), for US Army chain saw, World War II. This was part of the tool kit carried in many military vehicles, and was primarily intended for the disposal of trees felled as obstruction.

(Above and right) Entrenching tool adopted by US Army in 1943 to replace 1910 pattern shovel; blade folds for use as either spade or mattock; canvas case has suspension hooks to fit on field pack or ammunition belt. (Below) Japanese Army wire cutters, World War II; wooden handles shown unscrewed; canvas case.

Bibliography

There are very many books in print that provide reference for collectors of militaria. The following list is, therefore, not comprehensive, but every effort has been made to include those titles most likely to be of use to the novice collector.

P.E. Abbott & J.M. Tamplin
British Gallantry Awards (London, 1971)
J.R. Angolia
For Führer and Fatherland, Military, Civil and Political Awards of the Third Reich (2 vols, Stilwell, Kan., no date)
J. Atwood
The Daggers and Edged Weapons of Hitler's Germany (Berlin, 1965)
T. Baldwin
German Military Waistbelts (London, 1982)
A.J. Barker
Red Army Uniform (London, 1976)
Japanese Army Handbook 1939-1945 (London, 1979)
R.M. Barnes
A History of the Regiments & Uniforms of the British Army (London, 1950)
Military Uniforms of Britain & the Empire, 1742 to the Present Time (London, 1960)
R.J. Bender & H.P. Taylor
Uniforms, Organization & History of the Waffen-SS (California, 1969)
C. Blair
Pistols of the World (London, 1968)
W.H. & K.D. Bloomer
Scottish Regimental Badges 1793-1971 (London, 1982)
R.J. Bragg & R. Turner
Para Badges & Insignia of the Canadian Airborne Forces (Blandford, UK, 1980)
J. Brinkmann
Orden und Ehrenzeichen des "Dritten Reiches" (Minden, no date)
J. Britton
Uniform Insignia of the US Military Forces (Tulsa, Okla., 1980)
E.L. Bucquoy
Les Uniformes de L'Armée Française (Paris, 1935)
J.M. Bueno
Uniformes del III Reich (Madrid, 1977)
P.H. Buss & A. Mollo
Hitler's Germanic Legions (London, 1978)
W.Y. Carman
Headdresses of the British Army, Yeomanry & Cavalry (2 vols, London, 1968-70)
J.A. Carter
Allied Bayonets of World War II (London, 1969)
J.A. Carter & J. Walter
The Bayonet (London, 1974)
H. Cole
Badges on Battledress (London, 1953)
Formation Badges of World War II (London, 1973)
M. Cooper
Uniforms of the Luftwaffe 1939-45 (London, no date)
R.H.W. Cox
Military Badges of the British Empire 1914-1918 (London, 1982)
H.P. Davies
British Parachute Forces 1940-45 (London, 1974)
B.L. Davis
German Army Uniforms & Insignia 1933-1945 (London, 1971)
German Parachute Forces 1935-1945 (London, 1974)
Luftwaffe Air Crews, Battle of Britain 1940 (London, 1974)
US Army Airborne Forces Europe 1942-1945 (London, 1974)
Flags & Standards of the Third Reich (London, 1975)
German Uniforms of the Third Reich (New York, 1980)
B.Dean
Helmets & Body Armor in Modern Warfare (USA, no date)
R. Dilley
Japanese Army Uniforms & Equipment 1939-1945 (London, 1970)
T.J. Edwards
Regimental Badges (London, 1980)
J. Gaylor
Military Badge Collecting (London, 1977)
L.L. Gordon
British Battles & Medals (London, 1971)
H.D. Götz
Die Deutschen Militärgewehre und Maschinenpistolen 1871-1945 (Stuttgart, 1974)
R.W. Gould
Campaign Medals of the British Army (London, 1982)
L. Greer & A. Harald
Flying Clothing (London, no date)
G. Grosvenor (et al)
Insignia & Decorations of the US Armed Forces (Washington, DC, 1945)
J. Harrel
Regimental Steins of the Bavarian and Imperial German Armies (Wurzburg, 1971)
J.E. Hicks
French Military Weapons (London, 1964)
P. Hieronymussen
Orders, Medals & Decorations of Britain & Europe in Colour (London, 1967)
E.J. Hoffschmidt
Germany Army Uniforms & Insignia 1871-1918 (Connecticut, 1972)
I.V. Hogg & J. Weeks
Pistols of the World (London, 1978)
G. Hughes & C.A. Fox
A Compendium of British & German Regimental Marks (Brighton, 1975)
D. Jarret
British Naval Dress (London, 1960)
D.E. Johnson
Collector's Guide to Militaria (London, 1976)
B. Jones & B. Howell
Popular Arts of the First World War (London, 1972)
E.C. Joslin
The Standard Catalogue of British Orders, Decorations and Medals with Valuations (2 vols, London, 1979-81)
D. Judd
Posters of World War II (London, 1972)
R. Kahl
Insignia, Decorations & Badges of the Third Reich (Netherlands, no date)
P. Kannik
Military Uniforms of the World in Colour (Blandford, UK, 1968)
E, Kerrigan
American War Medals & Decorations (New York, 1967; London, 1973)
American Badges & Insignia (New York, 1968; London, 1971)
A. Kipling & H. King
Headdress Badges of the British Army (2 vols, London, 1980)
K.G. Klietmann
Deutsche Auszeichnungen usw., Deutsches Reich 1871-1945 (Berlin, 1971)
J.K. Kube
Militaria der deutschen Kaiserzeit, Helme und Uniformen 1871-1914 (Munich, 1977)
H.A. Lawson
A History of the Uniforms of the British Army (5 vols, London, 1962-67)
B. Le Marec
Les Françaises Libres et leurs Emblèmes (Paris, 1964)
D. Littlejohn & C.M. Dodkins
Orders, Decorations, Medals & Badges of the Third Reich (London, 1963)
H.A. Maeurer
Military Edged Weapons of the World (New York, 1967)
V. Mannstein & W. Buxa
Die Deutsche Infanterie 1939-45, eine Dokumentation in Bildern (Bad Nauheim, 1967)
G. Markham
Japanese Infantry Weapons of World War Two (London, 1976)
P. Marton & G. Vedelago
Le Uniformi Tedesche della Seconda Guerra Mondiale (Milan, 1980)
W. May & W.Y. Carman
Badges & Insignia of the British Armed Services (London, 1974)
V. Mericka
Orders and Decorations (London, 1967)
A. Mollo
World Army Uniforms since 1939 (London, 1980)
Naval, Marine & Air Force Uniforms of World War II (Blandford, UK, 1975)
F. Myatt
Modern Small Arms (London, 1978)
The Illustrated Encyclopedia of Pistols and Revolvers (London, 1980)
T. Nakata
Imperial Japanese Uniforms & Equipment (Tokyo, 1973; London, 1975)
C. Narbeth
Collecting Military Medals (London, 1971)
A. North
An Introduction to European Swords (London, 1982)
H.L. Peterson
Daggers & Fighting Knives of the Western World (London, 1968)
J. Pia
Nazi Regalia (New York, 1971)
P. von Pietsch
Formations und Uniformierungsgeschichte des Preussichen Heeres 1808 bis 1914 (2 vols, Germany, 1963-66)
A. Purves
Collecting Medals & Decorations (London, 1968)
R.H. Rankin
The Illustrated History of Military Headdress 1660-1918 (London, 1976)
H. Ripley
Buttons of the British Army 1855-1970 (London, 1971)
J.C. Risk
British Orders & Decorations (London, 1973)
B. Robson
Swords of the British Army (London, 1975)
G. Rosignoli
Army Badges & Insignia of World War II (2 vols, Blandford, UK, 1972-75)
Army Badges & Insignia since 1945 (Blandford, UK, 1973)
Ribbons of Orders, Decorations & Medals (Blandford, UK, 1976)
Badges & Insignia of World War II, Air Force, Naval, Marine (London, 1980)

U. Schiers
Deutsche Helme 1897-1914 (Fridingen, Germany, 1978)
J.L. de Smet
Colour Guide to German Army Uniforms 1933-1945 (London, 1973)
Smith & Pelz
Shoulder Sleeve Insignia of the US Army (Evansville, Ind., 1978)
W.H.B. Smith
Book of Pistols & Revolvers (New Jersey, 1979)
S.L. Stanton
Vietnam Order of Battle (USA, no date)
F.J. Stephens
The Collector's Pictorial Book of Bayonets (London, 1971)
Fighting Knives (London, 1980)
E. Stockton & M. Charlton
Reproduction Nazi Insignia (London, 1971)
J. Sweder
US Army Special Forces Insignia (2 vols, USA, no date)
US Army Ranger, LRRP and Recon Insignia (USA, no date)
S.W. Sylvia & M.J. O'Donnell
Uniform, Weapons & Equipment of the World War II G.I. (London, 1981)
H. Taprell Dorling
Ribbons & Medals (London, 1974)
W.T. Thorburn
French Army Regiments & Uniforms (London, 1969)
J.C.Tily
The Uniforms of the United States Navy (New York, 1964)
F. Tubbs
Stahlhelm (USA, 1971)
J.D. Walter
The Sword & Bayonet Makers of Imperial Germany 1871-1918) (London, 1973)
R.A. Westlake
Collecting Metal Shoulder Titles (London, 1980)
F. Wilkinson
Edged Weapons (London, 1970)
Collecting Military Antiquities (London, 1976)
Badges of the British Army 1820-1960 (London, 1980)
J. Wilkinson Latham
British Military Swords from 1800 to the Present Day (London, 1966)
M. Windrow
World War 2 Combat Uniforms & Insignia (Cambridge, UK, 1971)
T.A. Wise
A Guide to Military Museums (Hemel Hempstead, UK, 1971)
R. Wohlfeil & H. Dollinger
Die Deutsche Reichswehr, Bilder, Dokumente Texte zur Geschichte des Hunderttausend-Mann-Heeres 1919-1933 (Frankfurt, 1972)
H. Woodend
British Rifles: Catalogue of the Enfield Pattern Room (London, 1981)
J. Zienert
Unsere Marineuniform (Hamburg, 1970)

Collectors' Societies

United Kingdom

Arms & Armour Society
Secretary: Mr Joseph G. Rosa
17 Woodville Gardens
Ruislip, Middlesex

Chute & Dagger (Para & Special Forces Insignia)
Secretary: Mr J. Barker
Flat 3, 120 Grenfell Road
Maidenhead
Berkshire SL6 1HD

Historical Breech-Loading Small Arms Society
Imperial War Museum
Lambeth Road
London SE1 6HZ

Military Heraldry Society
Secretary: Mr T.J. Sampson
47 North Road
Bristol BS6 5AD

Military Historical Society
Mr R. Westlake
Wyld Way
Wembley, Middlesex

National Rifle Association
Bisley Camp
Brookwood
Woking, Surrey GU24 0PB

Orders & Medals Research Society
Mr N.G. Gooding
11 Maresfield
Chepstow Road
Croydon CR0 5UA

Scottish Military Collectors Society
Secretary: Mr M.S. Davidson
Findon Croft
Findon, Portlethen
Adberdeen AB1 4RN, Scotland

United States of America

American Society of Military Insignia Collectors
Mr George Duell
526 Lafayette Avenue
Palmerton
Pennsylvania 18071

Imperial German Military Collectors Association
PO Box 651
Shawnee Mission
Kansas 66201

International Military Arms Society
Secretary: Mr Edward H. Converse
Route 1
Summerfield
Ohio 43788

Military Collectors Journal
Mr V. Luska
Box 4393
Allentown
Pennsylvania 18105

Military Historical Society
"Adjutants Call"
Mr Peter Blum, PO Box 39
Times Square, New York

Australia

Military Historical Society of Australia
262 Tucker Road
Ormond East SE 14
Victoria

German Federal Republic

Bund Deutscher Ordenssamler
President: Herr Werner Sauer
Eisenbergstrasse 10
Postfach 1244, 6497 Steinau

Deutsche Gesellschaft für Heereskunde
Secretary: Dr Joachim Niemeyer
Finkenweg 2
7550 Baden-Baden

Gesellschaft für Historische Waffen- und Kostümkunde
Jebenstrasse 2
1000 Berlin 12

Italy

Unione Nazionale Collezionnista d'Italia
"La Voce del Collezionnista"
Signor M. Fasparinette
Via Lattanzio 15a, Rome

Picture Credits

The publishers wish to thank all those individuals and organisations, listed below, who made available material for photography or supplied photographs for the introductory sections of this book. In the latter case, copyright photographs are credited by spread number, from left to right and from top to bottom. The following abbreviations have been used:

IWM — Imperial War Museum
NAM — National Army Museum
USA — US Army
USAF — US Air Force
USMC — US Marine Corps
USN — US Navy
USNA — US National Archives

Jacket, Pages 1-7: material made available by J. Lyndhurst, Warnham War Museum & M. Fisher, "Regimentals"; **8-9:** USA, USA, G. Gardiner, NAM; **10-11:** NAM, USAF, NAM; **12-13:** USA, IWM, NAM, NAM; **14-15:** USA, IWM, United Nations, USMC, USMC; **16-17:** G. Gardiner, Salamander Books, G. Gardiner, NAM; **18-19:** USA, IWM, USAF, NAM; **20-41:** material made available by G. Gardiner, E. Campion, D. Chester, T. Edmonds, P. Gasnier, K. Holmes, G. Lay, A. Morris, A. Todman; **42-43:** IWM, USNA, NAM, IWM; **44-45:** IWM, USNA; **46-55:** collection of B. Cayley, (Insets) Warmham War Museum; **56-57:** collection of Mrs Y. Lyndhurst; **58-59:** collection of Mrs V. Walker; **60-61:** collection of D. Warneck; **62-63:** collection of T. Bradley; **64-65:** collection of T. Bradley, (Inset) collection of D. Warneck; **66-67:** Wehrgeschichtliches Museum, Rastatt-Schloss, German Federal Republic; **68-69:** Salamander Books, USAF, NAM, NAM; **70-71:** IWM, Salamander Books, USN; **72-77:** material made available by J.D. Sheen; **78-79:** collection of D. Warneck; **80-83:** Wehrgeschichtliches Museum; **84-85:** USA, USMC, IWM; **86-91:** collection of Miss M.E. Nobbs; **92-93:** collection of T. Walker and Warnham War Museum; **94-95:** collections of T. Walker, Mrs V. Walker, Warnham War Museum, (Inset) collection of Miss M.E. Nobbs; **96-97:** collection of T. Bradley; **98-103:** Wehrgeschichtliches Museum; **104-105:** material made available by G. Gardiner and Warnham War Museum; **106-107:** Warnham War Museum; **108-109:** USNA, USA, IWM; **110-111:** G. Gardiner, IWM, G. Gardiner; **112-115:** collection of Miss S. March, (Insets) Warnham War Museum; **116-117:** collection of D. Warneck; **118-119:** collection of T. Walker; **120-121:** collection of T. Walker, (Inset) Warnham War Museum; **122-123:** collection of T. Walker, (Inset) Wehrgeschichtliches Museum; **124-125:** collection of T. Walker, (Inset) Warnham War Museum; **126-127:** collection of T. Walker (Inset) Wehrgeschichtliches Museum; **128-129:** collection of T. Bradley; **130-137:** Wehrgeschichtliches Museum; **138-139:** Warnham War Museum; **140-141:** USMC NAM, IWM, USA; **142-143:** USNA, USA, NAM; **144-155:** The Weapons Museum, School of Infantry, Warminster, (Insets) Wehrgeschichtliches Museum; **156-157:** collection of M. Priest; **158-159:** collection of M. Priest, (Inset) Weapons Museum, School of Infantry; **160-161:** collection of M. Priest; **162-163:** USNA, Wehrgeschichtliches Museum, Wehrgeschichtliches Museum; **164-165:** collection of T. Gander; **166-167:** Warnham War Museum; **168-169:** Warnham War Museum, (Insets) collection of Mrs V. Walker and Warnham War Museum; **170-173:** Warnham War Museum; **174-175:** Warnham War Museum, (Inset) collection of T. Walker; **176-177:** Warnham War Museum, (Inset) collection of T. Gander; **178-179:** Wehrgeschichtliches Museum; **180-181:** USMC, USA, USN, USA; **182-183:** Warnham War Museum; **184-185:** Wehrgeschichtliches Museum; **186-187:** Warnham War Museum and collection of M. Little, (Inset) collection of Mrs. V. Walker; **188-189:** Warnham War Museum and collection of Mrs. V. Walker, (Inset) collection of Mrs. V. Walker; **190-191:** Warnham War Museum; **192-193:** Warnham War Museum, (Inset) Wehrgeschichtliches Museum; **194-201:** Warnham War Museum; **202-203:** Warnham War Museum, (Inset) collection of D. Warneck; **204-205:** Warnham War Museum.

PRINTED IN BELGIUM BY
proost
INTERNATIONAL BOOK PRODUCTION

POLAND